Time & Everything

Chain Loop Theory

Gerald Siegle

A Problem of Science or Fiction

TIME AND EVERYTHING

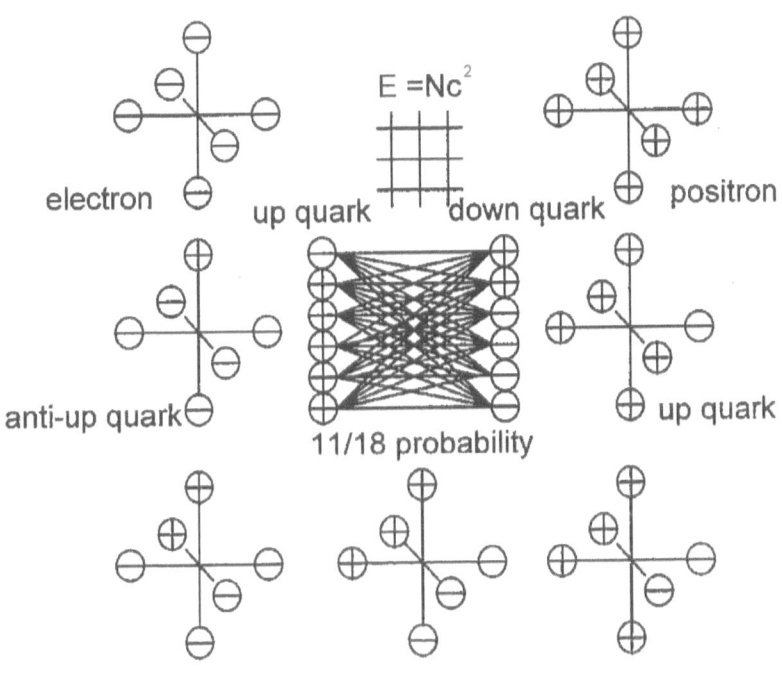

Universal Time-Leap-Loop Theory
&
Crossed Chain-Wave-Particle Theory

Library of Congress Control Number: 2010918673
ISBN: Hardcover   978-1-4568-3519-4
                 Softcover   978-1-4568-3518-7
                 Ebook   978-1-4568-3520-0

This book was printed in the United States of America.

To order additional copies of this book, contact:
Xlibris Corporation
1-888-795-4274
www.Xlibris.com
Orders@Xlibris.com
87607

Contents

Introduction

What you are about to read may not be of interest to you unless you have a basic knowledge of physical sciences and have an insatiable desire to know more than what is in physics books. For instance why do we have three dimensions and time? Is energy, time, and distance the same thing at a basic level? Is there a cycle time in which every point in the universe is recycled one unit from its present position? Are waves propagated through a fluid aether point loop path? Is there a basic connection between all the forces in the universe? Why do quarks and leptons have the charges they do? What is a positive and negative charge? What is electro-magnetism? Why is everything in the universe so interconnected? Why does the EPR experiments give the wrong answer. And there are other questions science hasn't explained. If you read this simple document with an open mind you will be astounded.

Absolute Time

@

Absolute Distance

A point has no dimensions, so points cannot make a line, a plane or a cube: adding points results in just a point. Mathematical proofs show that there are as many point locations in an inch as there is in a distance of say a million miles. However, there is twice as many point locations in two inches as in an inch. Einstein believed that we live in a time-line-continuum and he couldn't accept quantum time. However, we know time exists and seems to be the basis of everything. Nothing can move without time and time is movement. And every movement in the universe is choreographed in time, like the speed of light. It seems that at the basic level of dimensions—time, space and distance are one. A leap in time is also a leap in distance and a leap in energy.

It seems that a time-line-continuum must exist and is the first dimension. The second dimension would be quantum leaps in the time-line-continuum. For instance an hour or second is defined by a number of much smaller quantum leaps in the time-line continuum. Of course there is more quantum time leaps in an hour than in a sec. A time-leap is just one of an infinite number of identities in a time-line-continuum. For instance look at the line_____there is an infinite number of point identifications in this line. We could choose a finite quantum-point-identity of time points, for instance . . . . and so on. Our world is one of the infinite number of quantum worlds on the infinite time-line. If an inch has an infinite number of points then 2 inches would have twice as many infinite number of points and so on.

10

Infinity

What is infinity? Quantum time makes mathematics work in our world . . . each quantum is discrete in size . . . $1 + 1 = 2$. $A = B$ means that any number, large or small will apply. $N/0$ means that no matter how many times you multiply 0 by itself it is always zero and will never reach a value N. The formula $X = y^2$ shows a parabolic curve on a Cartesian coordinate system. When X is 0, Y is also 0 . . . the curve moves down, passes through 0 and then curves up. If a line is made of an infinite number of points, [having no dimension], what drives the direction of the curve? It seems that the curve is driven by points having an infinitely small dimension [the point approaches zero but y/x is finite]. The baffling part of this is that it has length and this means it is greater than zero though very small. It's as though there is a link between something and nothing. However, a point doesn't have any dimension so it can't add up to anything but zero. Just what is distance? I'm going to accept the idea that we live in a quantum-time-line world fixed in a time-line-continuum and see what this means.

Our world is composed of a large number of leaps of precise time-length. The quantum—time-leap is an identity transfer from a past to present position, a flow of time is the basis of everything.

"Identity transfer" from point to adjacent point through a lattice has already been used to solve problems in QCD (Quantum Chromo Dynamics). Physicist Tsung-dao Lee and his staff found that energy transfers in a lattice system could, in some ways, better produce the calculations being done in the branch of quantum field theory known as QCD, than calculus can.

Maybe it is possible to further reduce the lattice theory. We are conscious of a linear flow of time [timeline]. It just so happens that a continuous series of time-point-locations automatically is more than linear . . . it is a combination of many numerical arrangements all existing at the same time. A series of point identity transfers is also a repeating sequence [identity groupings] such as 1-2-3, 1-2-3 . . . . and so on. An interesting thing about this particular chronological repeating series is that it can represent three-dimensional space. For instance let's label the first point identity transfer X., the second Y and the third Z. Each 1-2-3 has three independent identities, X-Y-Z, each having a linear position in the series. A vacillating path identity would exist as a special combination of x, y and z identities chosen from the X-Y-Z series. For instance take x as the start of a universal point loop path then choose y or z randomly to complete the looped universal point path. The universal point path is repeated so that each chronological point of the loop path is one unit from the former. Repeating sequences also exist for many dimension say 1-2-3-4-5 point sequences representing 5 dimensions; our universe has basically 3 dimensions; other dimensions like relativity are mechanical time relationships of light in moving objects that are constructed from the three basic dimensions.

The chain path will have a finite number of forward and reverse (x, y, z) locations; the path can't be infinite or the series couldn't repeat. The last location of the chain path must be one identity transfer from the first in the cycled path. The process is repeated, as each chronological point on the past path is relocated one point unit from the past universal point path to the present point path. Cosmic cycle time is the time it takes to recycle all the points in the simulated universal looped point path—two present points cannot occupy the same position in the time cycle. Selection of the next relocation cannot include the occupied position of the present path and results in relocation away from the occupied point, I will call this a Blocking Point [this is the cause of inertia]. All the above is a special system that I believe represents our universe.

An interesting thing about chronological order identity transfer is that every point has control over the relocation of all other points and basic particles are the same anywhere. Another interesting thing is that a three-dimensional Cartesian co-ordinate system can be simulated by an interval of time quanta.

Rules for generating a universal point loop system

1) There are six possible point identity relocation directions, (+ or -) x, y, z.

2) All relocations, of a point., must be one unit from its chronological partner. For instance the last point, of the loop path, becomes the present point, but is one unit away from the first to complete the loop. Each of the past set of points are chronologically moved one unit to the present then randomly moved one unit to the future and so on as the future becomes the present identity each cycle time.

3) As each chronological present point is relocated its past chronological position ceases to exist as an identity of the system. In other words only the present special Point loop system is relevant or remembered from all possible point systems as present in cosmic cycle time.

4) A relocating point cannot occupy a spot already taken by another present point on the path and must relocate to another open location a slot from the filled one; I call this a blocking point meaning it is a point that blocks a relocation of a present relocating point.

5) There is a (+ or -) value for each of the x, y and z point location along the loop. A + value represents a forward

repositioning of a point while a minus means a backward relocation of a point. Minus values are added to plus values of point relocations along the loop from start to recycle will be one unit from the next cycle. In other words the end point of the cycle will be one point away from the chronological recycle point on the point loop path.

6)   The chart below shows a mini-co-ordinate system and its "remembered" locations. I use the word remembered, but I infer a special universal loop identity from all those that exist outside the chosen universe. We observe by being part of the system we are in.

Time   Points on universal loop path

--------------	(x	. . . . y	z)	
0	(0	0	0)	Starting point in this case is 0.
1	(0	1	0)	Move one unit in the y direction. Others are fixed in position.
2	(1	1	0)	Move in the x direction.
3	(1	1	1)	Move in the z direction.
4	(1	0	1)	Move in the -y direction.

Note: the rules are the same as nature when moving from one direction to another, just add or subtract a transfer from one of the three possible directions.

In a time-leap system there isn't any diagonals . . . just abrupt changes in direction [X,Y,Z]. Locations are in the time-leaps in one of the three groups. The time-line continuum has no beginning or end and always has existed [Actually I have no idea why time is there in the first place.] The first dimension is a time-line continuum. Quantum point locations in the continuum are a leap equivalent in time, energy and distance the third dimension is a division of the linear time-line-continuum into 3 groups (x, y and z) forming a separate identity path from the linear time-line-continuum. The fourth dimension is the back and forth recycling of the path, one point from the start in chronological order to recycle points in a secondary identity in time. A fifth dimension occurs when waves are propagated across impacting chain segments to form a crossed chain wave particle.

You will be startled by how this model evolves into a world like ours. No need for a man-god to create anything . . . with all the havoc the myth has put on the world. The universal-time-leap is interesting because an infinite number of universes are possible . . . just change the depth of the quantum-time-leap in the time-line-continuum. An infinite number of world divisions are possible . . .

Some paths are dead ends . . . while others are perpetual like the wiggling square of figure I to figure 2. Every choice is possible in an infinite world of possible relocations of a path [all paths exist]. The choices are identities as parts of a whole system., and always exist infinitely. There is no beginning or end . . . unlike film in a motion picture camera. Each time exposure is slightly different . . . a dynamic system.

Time, distance and energy are the same thing . . . one universal leap in the time-line continuum path. It seems to me that everything in nature obeys mathematical rules and is not spiritual or irrational. Everything we see is a deception—for instance in relativity we see time as constant, but instead it is not so—a person traveling through the aether would have time slowed down while a person fixed in the aether would age faster. We are also deceived by lengths as we are in time.

If we are deceived by nature, as in relativity, then maybe we are deceived by assuming that 3-dimension are not linear. Aether has to start with numbers and the simplest thing is a point and a relocation. A point has no dimension—a time-line is linear and x, y, z linear groupings are the aether that gives our impressions of propagated waves and gives movement and three dimensions, while empty space is nothing. Basically the universe is a mechanical, mathematical identity of time-leaps that can move backwards or forwards in cycle time.

A point location in space is a real thing—a time-leap is an identity transfer. We see objects moving, but maybe we just see time-line leaps that recycle and matter really is just the identity of points relocated in time.

Unversal Point Looped Path
The Fourth Dimension

Example 1

Even though time moves forward the point locations in linear-time will appear to have three dimensions and the path will assume a square shape. The time-line-continuum is first divided by 3 time-leaps in group X , then divided by 3 time-leaps in the Y group and then 2 time leaps are reversed from group X. The drawing shows a square path and movement of time-leaps from start to finish. The path is shown by a row of location figures x—y—z . . . . in time.

Because the square is fixed in two dimensions it's square form will be fixed if the one on the y line [a point unit from start] will fall in-to the start space and each point on the square will fall back counterclockwise . . . the finish of the cycle will then be on 2x.

How to change the shape of the square to a rectangle
Resulting in the ability of space to compact

I'd like to demonstrate how a 3-dimensional stepped loop path can squeeze space or relax space using the example of transforming a square path into a rectangle path having the same number of time-leaps. Starting with . . . .

Fig 1A . . . . 7 quantum time leaps in a time-line continuum are separated between 2 groups [x and y]

The figure is a square.

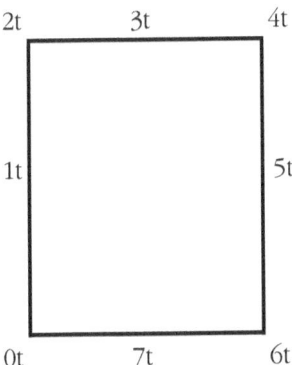

Fig. 1B . . . . 7 quantum time leaps in a time-line continuum are separated between 2 groups [x and y].

The figure is a rectangle, all points are one unit from the starting point [figure A]

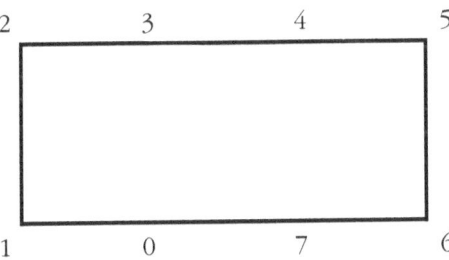

This simple example shows how the rules for a point-time-leap can change shape. All the time related time-leaps have relocated clockwise from the first cycle to the second cycle. The starting point 0 has moved to the right as shown.

The locations for 0 [start] is now x = 1 leap. 1 is now 0. 2 is now y = 1s leap . . . and so on

Fig. 1C

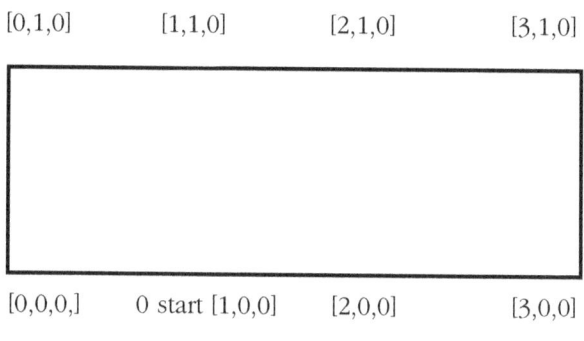

[0,1,0]   [1,1,0]   [2,1,0]   [3,1,0]

[0,0,0,]   0 start [1,0,0]   [2,0,0]   [3,0,0]

| | 1t | 2t | 3t | 4t | 5t | 6t | 7t | 8t |

First cycle [fig A] Start [0, 0, 0] ... [0, 1, 0] ... [0, 2, 0] ... [1, 2, 0] ... [2, 2, 0] ... [2, 1, 0] ... [2, 0, 0] ... [1, 0, 0] end of 1st cycle.

Second cycle [fig B] [1, 0, 0] ... [0, 0, 0] ... [0, I, 0] ... (1, I, 0] ... [2, I, 0] ... 3, 1, 0] ... 3, 0, 0] ... [2, I, 0] two end of series.

There are 7 time-points in the system from the starting point, the 8th relocation is the start of the next cycle. The last relocation is one unit from the first the next relocations will be one unit from their chronological past point.

Random choices can make different shapes of the path . . . 3-dimensions are much more complicated, but a path in 3-d is done the same way as 2-d. Our world has zillions upon zillions of points in a universe cycle.

Types of transfers

1) **Inline transfers** follow the point path so that the present location is one unit from its past point location. A tight ball of chains having only inline identity transfers will not change shape. Therefore inline transfers freeze time.

2) **Directional transfers** occur when all transfer points of a large segment of the universal chain loop move in the same direction. No internal movement and time is present as the object moves through space. For instance if all points in a clock move in one direction the hands would not move.

3) **Random transfers** are transfers that are not in-line down the universal point path. They are dynamic and can cause movement because random choices produce a random path that constantly changes shape All transfers will be randomly chosen unless only in-line identity are possible.

4) **Combination in-line and random transfers** generate a dynamic system with transient or fixed pockets of frozen time as in our universe.

5) **Gravity black holes, dark matter and dark energy** occur when chain segments are compacted causing more in-line identity choices due to an increase in blocking points BPs.

6) Stretched aether chains in an expanded big bang will also cause in-line transfers that slow time. Loose space chains speed time due to less in-line transfers.

Fig 2

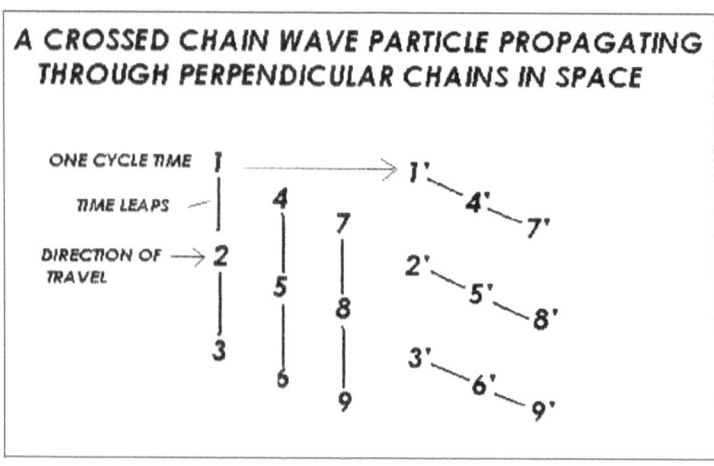

When the looped path is balled like twin, linear waves can form . . . for instance a chain moving in the Z direction will first strike an X-aligned chain of time-leaps, then strike Y aligned chains of time-leaps forming a crossed chain wave particle.

A linear wave forms when several chains in the aether path vibrate in unison to produce a forward biased wave and a reversed biased wave at the same time . . . each will move in opposite directions. The chains, in a forward biased crossed chain wave particle, move forward to strike perpendicularly aligned chains in it's path to propagate . . . . reverse biased chains fall from the aether into a chain void, as the void [crossed chain wave particle] moves away from the pressure particle much like moving your hand through air . . . there will be a pressure on the forward side of the hand and a vacuum on the trailing side of the hand.

A pressure space wave and a vacuum space wave (+ and - charge) are made by chance when random movements of chains, that are close together, move in a single direction . . .

A wave of crossed chains will form when traveling through a universal sized balled-loop-path of space chains because all the chain are forced to align to X, Y, Z space by the propagating wave. A wave traveling in the x-direction will strike y or z perpendicular aether chains in its path, likewise for y or z traveling waves. Funneling is due to propagation through aether chains, trailing chains form a net that funnels directionally relocating waves into particles. Propagation results in a crossed chain wave particle having rows = to points/row. In other words squaring the number of points on a row will give the total number of points in a crossed chain wave particle.

A forward biased wave will have a pressure front of space chains arranged close together and a trailing area of loosely confined space chains, while a reversed biased ccwp will have a frontal area of space chains falling into a void with compaction back to ambient space. A reversed ccwp is just a forward biased ccwp in reverse. A pressure particle will form a wake of space chain segments while a vacuum particle will form a wake in the opposite direction. These two types [+ & -] are just a probability wave formed from the time-line space chain path and after a short life span depart from their linear motion to random space vibrations.

A chain that is drawn tight is the strongest thing in the universe . . . it can't be bent except in stages at the ends. The stringiness of space explains the equal number of space charges. The more rows, the slower it goes . . . meaning that the velocity is inversely proportional to the number of rows. A point on the row is moved one time-leap per cycle through the universal point loop system. Each row resists movement . . . this is one source of mass. Both tension on the time-line segments and compaction of segments results in retardation of movement in other words mass.

Fig 3

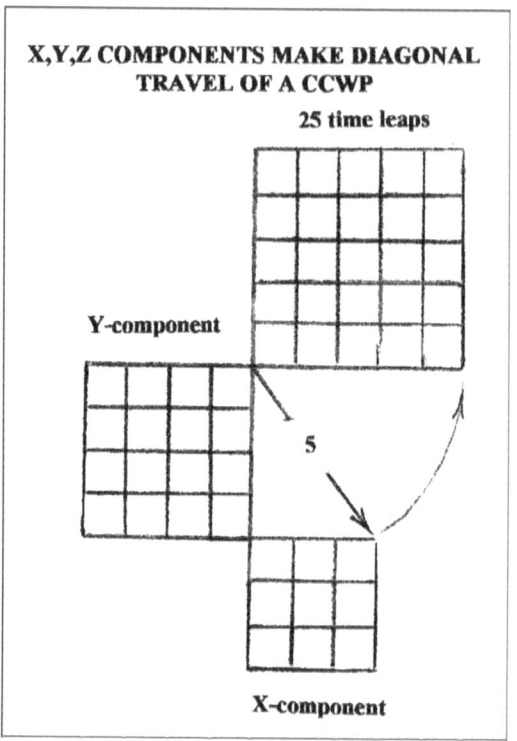

*the example for item 3) above take 25 steps in time units to move 5 step units in the given direction. The velocity will be 1/5 for a chain of a 25 time unit ccwp so the whole ccwp will have a 1/5 velocity in 25 step units of time. The result is 25/ 5 = 5. A 16x component will be 16/4 = 4 and a 9y component will have 9/3 = 3. For every 4 x-component movements the y-component will move 3 units. A 3..4..5 triangle is formed that has a velocity of 5 step units/ 25 units of time.

Characteristics of a crossed chain wave particle

1) Ccwp can only travel back and forth in three perpendicular directions.

2) Forward and reversed biased ccwp will have as many rows as there are points per row because transferred to perpendicular chains the points would divide between these chains so that there would be N rows each with N points.

3) The speed of a ccwp is dependant on the number of points/chain in the ccwp. The longer the chain the slower the speed. Since each chain has the same speed the speed of a ccwp is inversely proportional to the number of points/chain, $V = 1/p$. A point on a row of a ccwp is moved forward each cycle time of the universal point loop.

4) There are two types of ccwp, one is forward biased and the other is reverse biased. The chains in a forward biased ccwp lose their speed and drop back to ambient aether space as the ccwp moves forward. Reversed biased ccwp will have a leading edge into which space chains fall into the void and compress space chains into randomly vibrating space chains.

5) Forward or reverse biased ccwp are generated from random aether chain vibrations that by chance propagate a vector wave. They fill space and have various sizes and speeds, constantly being produce or canceled into vibrating space chain aether. They also have various sizes.

6) A forward biased ccwp is a pressure particle [a positive charge] while a reversed biased ccwp is a vacuum particle [a negative charge]. Positive and negative particles cancel into

vibrating aether chain vibrations . . . similar to hot and cold water poured together.

7) Negative and positive particles randomly occur when vibrating aether chains in space become synchronized to move in one direction . . . when this happens there will be a void and space chains will fall into this void as the void moves forward the void is a negative ccwp having taunt chains during propagation that cause more in-line identity transfers (mass).

Space filled with crossed chain wave particles.

Any where there is a time-loop-aether-system there will be crossed chain wave particles traveling in one of the 6 possible directions. Half of the ccwp will be positive in charge while the other half will be negative.

Like charges cancel, so the life expectance would be short. All the energy in our universe is in the time leaps of the random movement of the aether while matter and energy in our dimension are linear propagations in the aether. All virtual particles in space come from ccwp that fill space. A photon is a wave traveling through a field of ccwp, funneled by aether chains, while matter particles are centers of compacted space with charged fields. Electric fields are ccwp's [+ or-] that fill space, but the compacted core of a matter particle doesn't have a charge from the space field due to cancellations between [+ and -] charges, also fluid dynamics dictates that like particles will experience a vacuum between them that draws them together while oppositely charged particles are forced away from each other. This results in a single charge at each core surface instead of a mixed charge.

Fig 4

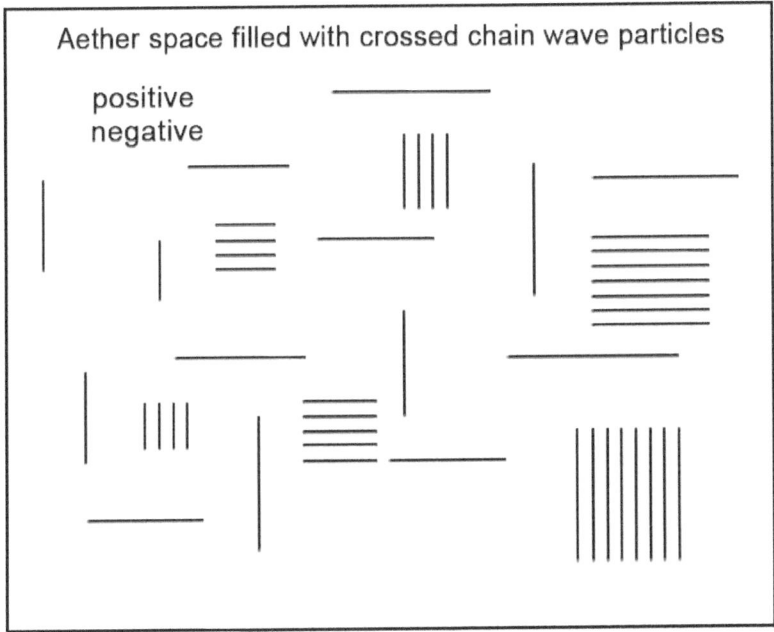

Aether space filled with crossed chain wave particles

positive
negative

If real space is empty it will have no dimension of any kind. Dimensions are part of the chain aether loop system and simulate directional movement and time. The chain aether loop system is a type of fluid aether. At the basic level, space, time and energy are the same thing. Space, time and energy diverge when crossed chain waves occur. Mass in a ccwp occurs when it is enlarged causing speed to slow down.

Crossed chain wave particles fill space; each crossed chain wave can move only in one of 3 perpendicular directions back and forth. An average sized crossed chain wave particle will travel at the speed of light and is the basic Plank quanta composing a photon. In other words many Plank particles or crossed chain wave particles compose a photon. Space is filled with all possible sizes of ccwp, the sizes [time-leaps] of ccwp would be 1,4, 9,16,

25, 36 . . . . N^2. All the energy of our universe is in the chain aether. Most of the energy in the universe is in random vibrations. At times these local vibrations, by chance, vibrate in one direction moving a short chain length that strikes other aether chains in its path and propagates crossed chain wave particles. As an example, if the starting chain has 9 energy points, it will divide between the other two dimensions to make 3 energy rows of three rows.

There are two types of ccwp, one is positive and the other is negative. Space is filled equally with both, Positive and negative crossed chain wave particles that annihilate each other as they are created from random vibrations, forming vector waves of crossed chain wave particles in an equilibrium.

A secondary wave moving through the ccwp field will confine all the N^2 ccwp's in a small Plank area by impacted aether chains that form a funnel. In a Plank area there will be ccwp of several sizes 1, 4, 9, 16, 25, 36 . . . N^2. A plank quanta will be N^2 half the number of leaps in the series of N square, or $N^2/2$ is the average ccwp particle that moves at the speed of light. A plank quanta can now travel in any direction at the speed of light because all the proper components are in place. $C^2 = X^2 + Y^2$. The wave will have a (+ or -) charge and will cancel the opposite charge as it propagates through space leaving the original charge and wave. After the wave has traveled through and canceled the + and - ccwp's, a short time is needed for a probability in the aether to rebuild + and - ccwp's in an equalibrium.

Several Plank particles bound together make photons of which the frequency is inversely related to the number of Plank particles.

Waves in air are compressible and that's why there is a sound wave. A single wave has a compressed part and a rarefied part giving a crest and a wave depression, the distance between the two is a wavelength.

Light is a boson and bosons are not supposed to bounce off one another, but travel through one another like sound waves do. However, if bosons are like smoke rings, can smoke rings travel through one another without changing their shape? We do know that bosons can change into fermions. Femions do rebound and for some reason and also collect bosons—how is this done? Then there are the fermion ghost particles that come and go. There seems to be a sea of waves in a looped time-leap system that is balled into a universe and waves form from a mix of small waves into larger waves which smash into one another and collect to form all the particles in the universe. Compaction by gravity in stars form elements that make our world.

Electromagnetism

The best explanation for electrodynamics and spin is a fluid aether vortex, Hakan Egne from Dokuz University in Turkey has demonstrated how this can be done using fluid dynamics to construct a vortex including the circulation of the fluid outside the vortex. The entire vortex is a donut shaped identity of circulating areas of ccwp's in the vortex and around it. A photon is composed of two parts, one counter rotating from the other. The two parts of the photon are forced together by the vacuum created by the counter rotation between the two particles as in fluid dynamics . . . the force of the vacuum is called magnetism. Magnetism and electrostatic force influence one another when stationary and the other is moving. The electrostatic force and magnetic force occur at the same time at the core of quarks and leptons and also in bosons.

In ccwp theory two sides of the core of quarks and leptons are driven together by probability waves in an x, y, z sequence. The x-y-z salient movement or 3-phases, swirl fields into vortexes causing electrons to follow the x-y-z salient movement. The driving energy comes from chance random to linear movement from the time-leap aether of space which diffuses into the core generating ambient space. This powers everything and is the source of everything—in this way no energy is ever lost, it is created in space by chance and returned to ambient space at the core of quarks and leptons.

Fluid dynamics shows how the two spin parts of a photon [in ccwp theory fields of - ccwp's and + ccwp's cause opposite X, Y, Z spins that evolve into an electron and positron at high energies. Also, the orbital of atoms become reasonable with fluid dynamics. A combination of the two theories unifies all physics.

Magnetism can be caused by moving a charge through space. Cancellations of opposite charges at the surface of the core of leptons and quarks is the cause of the vortex or spin as ccwp's move in and bounce off the surface of the core. Since ccwp's propagate at the speed of light through an aether the reversed spin is canceled while the forward spin is strengthened resulting in one spin. This happens with a charged particle such as an electron. The energy for the vortex comes from the cancellation of charges in space around particle cores. The shirality of space is due to x,y,z phase spins of canceling ccwp's.

Waves travel in a straight line—how could a wave that travels in a straight line form a vortex? In ccwp theory waves can travel only in x, y or z directions—however, a Planck volume of space would contain all the x-y-z components necessary to propagate light in any direction at a constant speed. The components are divided up by 1^2—2^2—3^2—4^2 . . . . N^2 . . . and so on. These components are balanced by oppositely moving components or the aether will travel. A photon composed of these (+ & -) components, in space, will cancel components in space leaving the forward moving photon to continue this process allowing the photon to travel at a constant speed in that direction. The cancelled components return to random movements of the space chains of the aether. Remember that the components of a photon conform to the Pythagorean theorem $D = (x^2 + y^2 + z^2)^{1/2}$. For a photon to turn it must be struck by another small particle to turn it or the density of space is changed from one local to another.

The energy of the three exponents, $x^2 + y^2 + z^2$ are also in the spin of a quark or lepton even when the object is traveling along an axis of space. A magnetic field happens when an electron travels in a circle and the magnetic field causes a vortex field of + and - ccwp waves that follow the electron.

In space, linear waves are generated by chance by random movements to linear movements of the time-line foundation of the aether. Moving the aether along an axis of space shifts the direction of time leaps to the vector direction traveled—relative time slows down in an object moving in the aether just as if it would from a fixed location in absolute space.

Special relativity

I'll give my version of relativity so anyone can understand it. To start with a light source will radiate light at the same speed in any direction—relative to absolute space. Think of a box car moving down a track with a boy throwing a ball at each end. To the boys the ball is traveling a given speed c. To the person standing on the track the speed of the ball moving away in the box car is v + c while in the return the ball has a speed of v - c. Both observers are observing the same ball—or in relativity of light, the same photon is travel across the boxcar from A to B.

Now think of a moving vertically aligned yardstick with a light wave traveling up the yardstick. The actual distance the light wave traveled is the diagonal C. The actual speed up the yardstick is shown in the drawing. The diagonal distance (speed) would be traveled by a light source on the bottom of the yardstick.

Fig 5

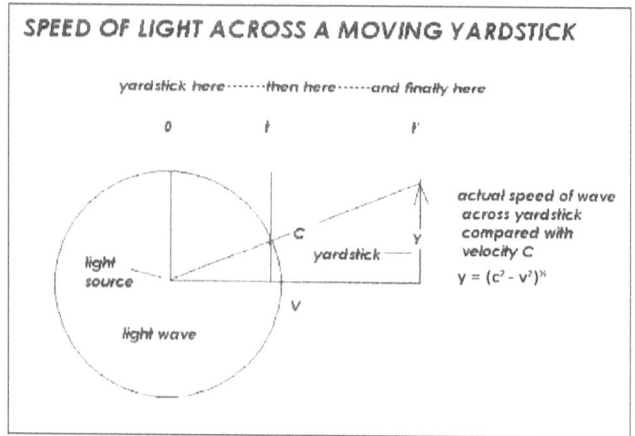

SPEED OF LIGHT ACROSS A MOVING YARDSTICK

yardstick here ·······then here ······and finally here

actual speed of wave across yardstick compared with velocity C

$y = (c^2 - v^2)^{\frac{1}{2}}$

Light traveling up the yardstick will leave a diagonal projection shown against a stationary system. To the stationary observer the sound wave is traveling speed C and slower up and down the yardstick (think of the a photon as the sound wave in the drawing and you have the basics of relativity.)

Photons are the basis of all matter fields which oscillate back and forth at the speed of light. Point A to B (the length of the boxcar) has not changed, only the relative speed of the ball has as it moves up the yardstick. The hands of a clock composed of photon fields runs slower. Everything happens slower in a moving system relative to a fixed position in space.

Think of the hands of a clock rotating at a constant speed and then driving the clock forward with force particles like photons that have energy. The photons are absorbed like snow balls thrown at a snowman in space—the snowballs will drive the snowman forward through space. Of course the mass of the snow man will increase with the addition of the snowballs. Relativity states that nothing can move faster that the speed of light relative to fixed space. To make an object move forward oscillations at the speed of light relative to fixed space must be vector orientated to travel in a direction. This is also a reason for mass.

In Figure 6 the relative speed across the boxcar from start point A to end point B is $(c^2 - v^2)^{1/2}$ from the Pythagorean formula for diagonals. The diagonal distance shown is travel time for a photon of light c in absolute space.

The boxcar travels at velocity v relative to the stationary observer. The speed of light in the moving boxcar would be $1/(c^2 - v^2)^{1/2}$ relative to c the speed of travel across the diagonal.—$c/(c^2 - v^2)^{1/2}$ and this reduces down to $1/(1 - v^2/c^2)^{1/2}$. When the diagonal is 1 unit the shrinkage in speed is $(1 - v^2/c^2)^{1/2}$.

Fig 6

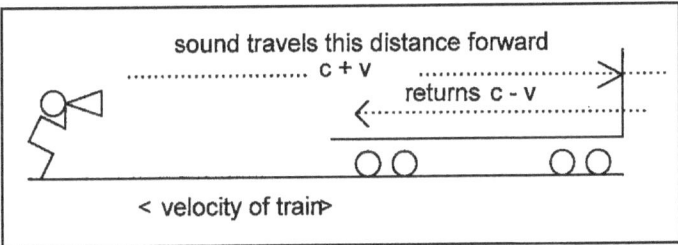

sound travels this distance forward
.... c + v .... ⟶ ......
⟵................................... returns c - v
OO   OO
< velocity of train▷

Fig 7

Michelson & Moreley's rotating mirror experiment

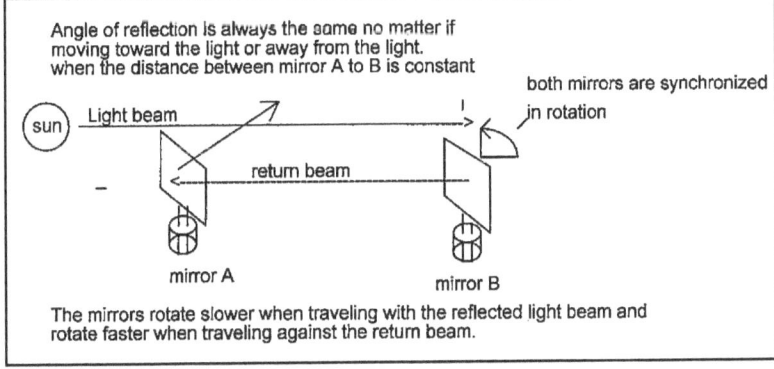

Angle of reflection is always the same no matter if moving toward the light or away from the light. when the distance between mirror A to B is constant

both mirrors are synchronized in rotation

sun Light beam

return beam

mirror A   mirror B

The mirrors rotate slower when traveling with the reflected light beam and rotate faster when traveling against the return beam.

The rotating mirror experiment above shows that time slows when moving relative to the aether; even clocks and rotating mirrors. For instance the distance between mirror A and mirror B is 10 miles, both mirrors would rotate a small angle @ in the time it takes to travel between them. The arc or angle would be the same if moving towards or against the source of light, relative to absolute space, even if traveling near the speed of light and it may take a million years to cross from mirror A to mirror B.

The mirrors rotation is diluted by time-leaps driving it forward and this dilution slows the actual time of rotation. When the mirrors on the earth are traveling towards the light source these time-leaps are removed and the mirrors speed up to the actual rotation speed relative to fixed space.

Light is a wave that propagates through a crossed chain wave particle aether, so the speed is the same as the sound wave in the drawing. In crossed chain wave theory, light behaves like the sound wave but at the sub atomic level can in some conditions have a slower speed after bouncing back relative to the moving observer. Crossed chain wave theory would fit in an earlier stage of evolution that makes the photon field that could move like wind. However, unlike wind, aether would experience relativity due to the stepped nature of relocations in the aether. Ccwp theory constitutes absolute speed in absolute space, while relativity is a function of ccwp's. If all time-leaps are forward in the aether (vastly faster than the speed of light) all movement, relative to the aether could not exist—nothing can move faster than the time-leaps.

Strangely though, ccwp theory gives the same speed of light when moving as does relativity, the relocation is in the same place for both views so they are the same thing , the vertical shrinkage in relativity is $1/(1 - v^2/c^2)^{1/2}$. While the back and forth movement is $1/(1 - v^2/c^2)$ or a $1/(1 - v^2/c^2)^{1/2}$ ratio between the two figures and equals the figure represented by the Lorentz-Fitzgerald transformations. I could explain the above this way, C+V + c-v = 2c for both directions or c in one direction. Considering that the diagonal distance is c the formula becomes $c/(c + v) + (c - v)]$ which is equal to $1/(1 - v^2/c^2)$. Crossed chain wave theory gives the same distance in time as relativity does for the same reason. For instance in crossed chain wave particle theory, light traveling across a yardstick aligned to the direction of travel, will move relative to the end of the traveling yardstick at $1/(c + v)$ and return to the start position on the yardstick at . . . $1/(c - v)$. The total distance is . . . $2c/(c^2 - v^2)$ which is equal to $2/(1 - v^2/c^2)$ in two directions or in one direction $1/(1 - v^2/c^2)$.

Sense time runs slower when moving relative to the aether the speed of both the relative system and its absolute system give the same speed for light. In the drawing below I have divided both sides of the relative movement $1/(1 - v^2/c^2)^{1/2}$ to give a crossed chain wave movement. The ratio is the same value but is relative to the aether and not between relative movement of an object as in Einstein's special theory of relativity. In this way there isn't any "twin paradox" that plagues special relativity. Actually, a Planck volume of space contains a given number of n^2 ccwps no more and no less so fig 5 can't be but it gives the right answer for contraction of a moving yardstick.

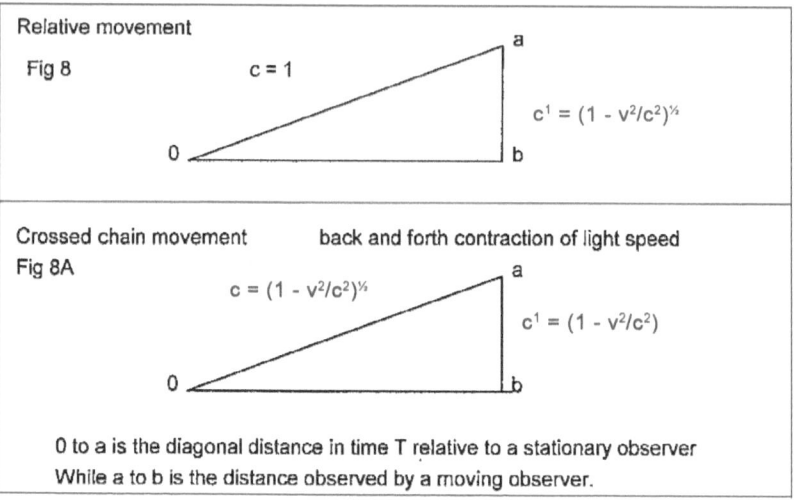

The relative effect is also identical to the crossed chain effect when considering that the vertical distance has shrunk by $1/(1 - v^2/c^2)^{1/2}$ and this means that for the diagonal to be 1 both sides would have to be multiplied by $1/(1 - v^2/c^2)^{1/2}$. Interestingly, the ratio between of shrinkage of actual speed of light traveling in the vertical directions and forward directions of light in the interferometer both are ratios are shown above.

Crossed chain wave theory and relativity

Below is the hypothetical size (not the real size) of a crossed chain wave particle seen from a moving object. The actual size of the ccwp, relative to fixed space, will be a single size. Relative to an object traveling away from the light a basic quantum of light will appear to be $(C - V)^2$. In this way it fits relativity as though velocity vectors dilute movement (time) in a moving object.

The formula for diagonal movement $D = (x^2 + y^2 + z^2)^{1/2}$ show crossed chain wave particles that move in 6-directions + or - (x, y and z). I am going to simplify the movement by setting x to 3 units of travel in 9 point units of time while y travels 4 point units in 16 units of time and I'll set z to zero. In this case the x-component moves 3 point units in 9 point units of time and then stops while the y-component takes over and moves 4 point units in 16 point units of time z is zero so that his cycle is finished and the next cycle starts. As in a 3-4-5 triangle 9/25 + 16/25 = 25/25. Note that the interrupted time results in 25 units of time for the x-component and also for the y-component. This results in a 3-4-5 triangle with a diagonal by multiplying both sides by 25. The diagonal line fits perfectly into relativity—no problem. 9/25 + 16/25 = 25/25 reduced to 9 + 16 = 25 and the velocity v is the square root of the point leap energy; so v = 5 in comparison to x and y.

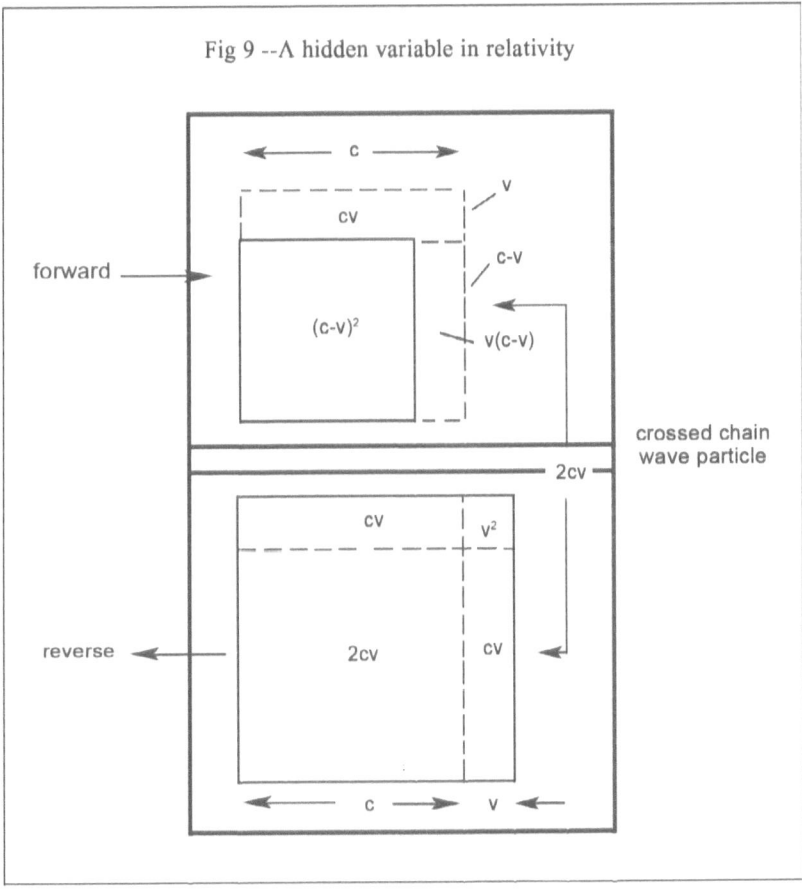

Fig 9 --A hidden variable in relativity

The drawing fig 9 shows a crossed chain wave particle at top that moves faster than the speed of light, hits a mirror and bounces back as a larger slower crossed chain wave particle. Here could be what's happening in a crossed chain wave particles in a Planck volume when it rebounds. First it travels at velocity c - v and then returns at velocity c - v leaving c the speed of light. Energy transfers are shown by the formulas below

$$(c - v)^2 = c^2 - 2cv + v^2$$

$$(c + v)^2 = c^2 + 2cv + v^2$$

2cv energy is added and subtracted from $c^2 + v^2$ in time-leaps while cv is momentum and appears on the forward and reverse directions. This concept is controversal and doesn't obey the the speed of light C in a medium—however, I added it in as it shares properties with x,y,z components of light in the movement of a wave at the same speed in any direction. A Planck volume contains + and - ccwp's that fill space and contain 1^2, 2^2, 3^2 ...N^2 number of x,y,z steps forward and backwards. The Planck volume can be divided into crossed chain wave particles that fill a Planck volume of space.

In the drawing c^2 is a standard sized ccwp, in a photon, in aether space. The $c^2 + v^2$ has 2cv energy (c is the number of leaps in a chain of a ccwp as mass) subtracted from it in the forward direction and the $c^2 + v^2$ has 2cv added to it in the reverse direction. Remember that this may not be right as the speed of light propagates relative to an aether. However, it may take part in movement of probability waves at sub atomic levels.

The number of time-leaps are the same for both the forward and reverse directions of a photon, meaning that there will be more smaller basic quanta in the forward photons and fewer but larger basic ccwp's in a photon, [the number of time-leaps in a photon are constant], but the speed and mass are different as the speed of a photon is inversely related to the size of its ccwp's and mass is related to the time-leaps in the photon.

Photon size represents it's speed relative to the aether. The addition and subtraction of velocities would only be in relation to the size of a photon relative to movement of the object. The missing points on the ccwp is the forward velocity points driving it—actually the size of the ccwp, composing the photons has not changed, but forward time-vectors have been diluting it.

To move an object forward it must be moved by driving it with force carriers such as photon which are absorbed and added to the mass. The formula would be the number of time-leaps in a chain of a ccwp diluted by forward leaps $T_{dil} = T_n/(T + T_n)$ $= n(1 + n)$ Time dilution is a ratio of the original time T compared to time-leaps added to the original time. This results in a slowing of rounds/time say for a rotating object that is accelerated, as in the rotating mirrors in the M & M experiment with light. Take away the forward vectors and the mirrors rotate at the speed it should when fixed in space.

If T_n number of leaps is added to the equation then the dilution is ½. That is if T_n is equal to the time related to the speed of light. There will never be enough steps to drive T_{dil} to infinity. As T_{dil} gets bigger the speed of the clock will run slower mass will be infinite to drive the particle to the speed of light.

An article from the web written by Roger Elhnan called, "A Reconsideration of Matter Waves" shows that using kinetic energy, $E = \frac{1}{2}mv^2$ used in DeBroglie's matter wave theory, the waves would move ½v the speed of the matter producing the matter waves. When using $E = mc^2$ the speed would be C^2/v, the wave moving faster than the particle producing the wave. The answer to the problem may lie in the action of the aether propagating the wave for instance the kinetic energy formula $E = \frac{1}{2}mv^2$ is derived from velocity from zero speed to v while $E = mc^2$ is derived from velocity zero to velocity 2v in which the velocity is already halved the value being v. From the article "Reconsideration or Matter Waves" (7) part of the formula on page 3 shows Vmw = (h/m-v) $(1/2mv^2/h)$ = 1/2v and the wave is traveling 1/2 the speed of the particle. Of course this is impossible unless the speed of propagation is 0 to 2v at a point traveled through the aether then v will be the velocity—the. speed is already halved. This means that for a wave traveling through space the kinetic energy formula for the wave would be $E = mc^2$. A sign wave takes a half circle moving upward then a half circle in the downward direction. The actual wavelength is twice the diameter.

Fig 5 shows that the speed of light is relative to each coordinate system no matter how fast the system is moving away. A person moving away from another will see his clock running slow, in fact each observer will see the other as aging slower and this is not the case. Relativity starts from a fixed position in absolute space and not between objects. This fault of special relativity is called the "twin paradox".

Lorentz transformations

Fig. 10

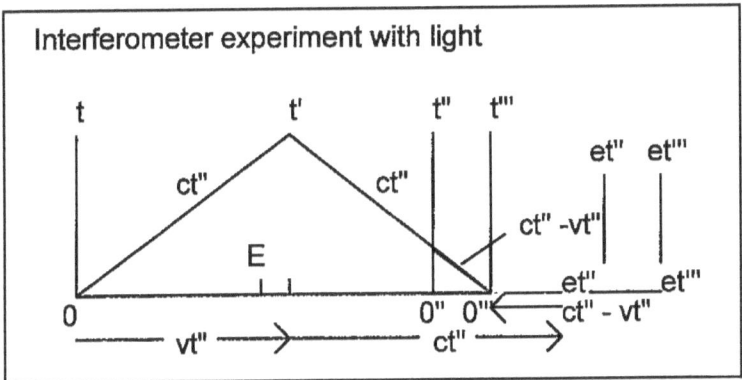

Interferometer experiment with light

The length 2C is the diagonal background distance of travel up and down the interferometer as its vertical arm moves forward. The background distance of travel for both the vertical and horizontal arm of the interferometer are equal. The distance between t' to t''' = et" to et'''. There is a shrinkage when the interferometer is at t' and the distance traveled in the horizontal direction in time t' shows a shrinkage.

The two lines, showing light travel, are equal in length and light will travel this distance at the same time. When moving towards light your speed and the speed of the light is added so that the returning light travels across the yardstick faster. The drawing isn't accurate but the distance 0" to 0''' is proportional to ct''' the length of the diagonal line ct'''. Also, 0" to et" is equal to t the length of the vertical distance 0" to f. The horizontal distance of the horizontal beam of the interferometer has not changed.

What is the meaning of $D = (x^2 + y^2 + z^2)^{\frac{1}{2}}$

The formula for diagonal distance through coordinated space as $D = (x^2 + y^2 + z^2 + t^2)^{\frac{1}{2}}$. D also can be thought of as a radius in a first quadrant like the picture below. In crossed chain wave particle theory there is another meaning to the projected diagonal line. A Planck volume 4.22419×10^{-105} meters—in crossed chain wave theory a Planck volume is composed of a lot of much smaller waves that have various speeds. For instance, for ccwp's the velocity is equal to the number of leaps across a chain of the ccwp by the formula $S = 1/L$ in which L is the number of time-leaps composing the chain. This means that the crossed chains in a Planck volume would be 1 leap, the next particle would be 4 leaps and the next would be 9 or . . . . L^2. To make a diagonal say $x = 3$ and $y = 4$ in a unit of time a ccwp having 9 leaps in the x-direction and 16 leaps in the y-direction would equal the same distance traveled as a 25 point crossed chain wave particle. Of course the distance traveled for a 25-leap crossed chain wave particle diagonal is a 3,4,5 triangle. In a Planck volume there will be for every ccwp a ccwp of the same size traveling in the opposite direction. A wave through Planck space would cancel reverse moving components and leave the other components of Planck space moving in the forward direction. The average speed of crossed chain wave particles in a Planck volume is their average speed and equals the speed of light.

Fig 11

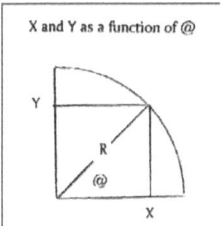

Using $r = (9 + 16)^{1/2}$, $= 5$ with $x = 3$ and $y = 4 - 25 = 9 + 16$. The projections on the x-axis is 1-2-3-4 equally spaced. 25 is the number of leaps necessary to move a distance of 5 units, 9 number of leaps are necessary to move 3 x-leaps.

Fig 12

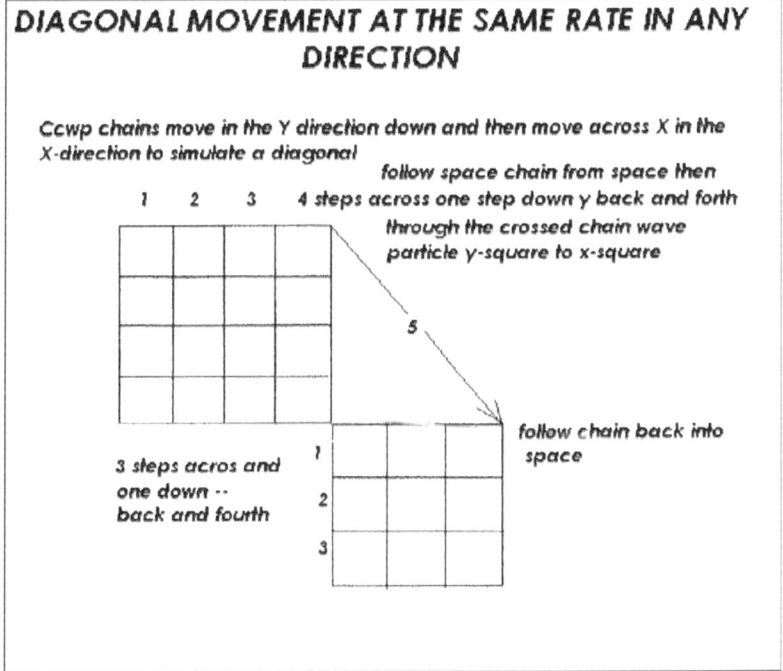

x-distance, then y-distance and then z-distance, at appropriate speeds, generating 3 dimensional groups of relocations. The stepped movement in three dimensions is part of the lattice theory used in science to simplify certain problems in physics. The most important aspect of ccwp-relativity is that it explains 3 dimensions as groupings of movements—just add or subtract from one of the three choices to relocate in a time-line.

If quantum time is 25n as in the formula $L = (x^2 + y^2 + z^2)^{1/2}$, then setting z to zero, $25n = 16xn + 9yn$. Of course this is a right triangle and distance covered by component $x = 4$ and $y = 3$ giving a diagonal of $L = 5$. The diagonal 5n is the length of travel in a fixed point in absolute space. [remember that 25n is the number of time-leaps in a time-line continuum. This means that for higher velocities the number of points in a ccwp is less as the formula $v = 1/L$ but the number 25n points is constant so there will be more smaller components] The number of point-leaps is constant for all systems in the universe.

Mass of a crossed chain wave particle

The formula for the velocity of a ccwp is $S = 1/L$, the reciprocal of the number of time-leaps across a chain of a ccwp. Sense the speed is the inverse of chain length then the chain length also expresses mass. Mass retards movement while velocity is the result of the retardation of velocity. From physics $F = m@$ adding time t to both sides gives $Ft = mv$. Also, the derivative of $E = mc^2$ is 2mc which is momentum 2mv. Isn't the derivative of E momentum mv?

Fig 5 shows that wave movement between two coordinate systems moving away from each other (relativity) can be simulated by alternating the sizes of outgoing ccwps and returning ccwps so that there is a velocity in the forward direction. This is shown by the formula $S_{fwd} = c + v$ and $S_{ret} = c - v$—squaring this gives the number of time-leaps totaled in a ccwp. The forward moving ccwp would be $c^2 - 2cv + v^2$ and a return ccwp would be $c^2 + 2c^2 + v^2$ due to $S = 1/L$. The value $2cv$ is subtracted from $c^2 + v^2$ for the forward movement and is gained by the return movement. The forward movement is greater and the return is less according to the speed of the ccwp leaving a velocity for the entire movement.

I notice $2cv$ is somehow related to the derivative of $E = mc^2$ which is $2mc$, momentum. Could $2cv$ also be momentum in which v is the mass? This would be momentum related to an object that is moving through space. If this is so then this explains why derivatives can give different dimensions such as suggested by Kuluza & Klien. Of course they are just an off shoot of our original 3-dimensions and time movements.

General Relativity and compaction of space

Spiral galaxies do not follow the restraints of Newton's mathematics . . . the stars at the outer edge are moving to fast. Ccwp theory states that a chain of a ccwp must move ahead one step/ universe cycle time. This mean that the chain will move forward at a constant speed and will take longer to reach a greater distance. Interference to relocation of points due to compaction means that the speed of light will be slower in an aether compaction state than in a less compacted state.

As the aether expands the speed of light increases and so does time. In the latter stages of expansion in-line identity transfers will play an insignificant part and distance and the speed of light will converge into a linear function. Stars in the outer fringes of spiral galaxy will travel faster than those at the center of the galaxy because there is less in-line identity transfers slowing time.

Sense energy, distance and time are the same thing, energy equals the total number of point-leaps in an area of C. In the energy formula $E = MC^2$, C is the speed of light as in the formula $v = 1/p$ . . . squaring this results in the total number of time-leaps in a photon. A basic Plank quanta is the energy unit in nature. All mater is energy composed of Plank quanta so all we need to know is how many Plank quanta there is in a mass and we will know the energy of the mass. This mass is in Einstein's energy formula $E = mc^2$. The speed of a Plank quanta is C, and $C = 1/p$, where p is the number of energy points in a row of a ccwp Plank quanta. The total energy E of a ccwp or Plank quanta is $E = C^2$.

Squaring both sides of of the equation $c^2 = 1/(p^2)$. However, the distance p occurs in p^2 time; this means the $1/p/p^2$ and this is equal to p. This means that $E = m/c^2$. Energy, space and time have separated from each other. Energy, space and time are the same in a single step; it will travel faster than the speed of light and the fastest speed possible as a wave propagating through space. This means that all the mass has been converted into energy. A Plank quanta has a given number of restrained (in-line identity transfers) energy points that cause it to travel slower than a single point ccwp.

This means that the Plank quanta has mass [resistance to movement] compared to a single point ccwp. We don't see this and believe nothing can travel faster than the speed of light so all the mass is in energy E, time T and distance D are therefore not the same thing in this case.

The speed of light $C = 1/p$ in T time. For instance if $p = 5$ and $T = 25$ the speed is 5/25 so $1/(5/25) = 5$ steps forward in 25 steps of time. The formula $E = mc^2$ in crossed chain wave theory is the number of c^2 crossed chain particles in a Planck volume of a photon.

Relationship between E =MC²
and E =1/2MV²

The kinetic formula $E = 1/2mv^2$ is the energy between zero to v. V/2 is the average speed when accelerating from zero to velocity V. The speed of light is already an average speed sense particles in the medium are accelerated from zero to top speed 2c, so $v/2(mv) = E$ when propagating the wave. Both formulas correspond to the same thing—$E = 1/2v(mv)$.

Look up "A Reconsideration of Matter Waves" by Roger Elhnan in the web. It shows a matter wave problem. DeBroglie proposed the idea in 1924 but the problem is now solved. The article starts by comparing the energy from special relativity $E = mc^2$ and the wave formula $E = hf$ where h is Planck's constant times f the frequency of the wave. This has experimentally been shown to be correct and shows a relation between wavelength in matter and the speed of light. Fig 9 on page 39 shows 2cv moved from one direction to the opposite direction. Momentum cv means that c is also mass and is the number of steps in a chain of a ccwp of c velocity. I wonder how this fits the conditions for velocity of the wave being = to the velocity of the matter wave.

The problem is that the speed of the wave given from one formula is 1/2v in another formula and this is impossible. The explanation I have given above seems to fit the bill. In the article mentioned the question is asked, "why is the formula $E = hf$ different from $E = mc^2$? The velocity v of the wave is v/2 of the traveling object.

The kinetic energy formula, of course, is wrong—it doesn't show the increase in mass (energy) as velocity is increased. It seems proper to have $E = MV^2$. As velocity v approaches c (the speed of light) it is total energy, because nothing can move faster than the speed of light. And this is the standard for energy conversion.

The reason that mass increases with speed, is because it takes a force field to move it forward and this addition of energy is caught and added to the mass of the moving object—this is another way of saying an object cannot move faster than light. Light is energy and there isn't anything faster than light to be thrown at an object to bring up its speed—also, matter is also a wave product. Special relativity states that time slows and with the slower time there is more mass. And therefore to push this increased mass, more force particles must be added to move it.

The derivative of $E = 1/2\ mv^2$ is momeutum mv. Momentum would be the value at the peak of acceleration not an average for v. The derivative of E in the formula $E = mc^2$ is 2mc. The mass of $E - mc^2$ is twice as great. It seams that the measurement for the speed c is an average between zero and the top speed of light 2c. This suggests that aether is compressed and rarefied to carry a wave. During propagation the photon aether is compressed from zero speed to 2c, the average is the speed of light.

If the specd of propagation through aether is twice that of light, it allows for a sign wave that would pass through a polarized lens. The most stable form of light would be a looped vortex like a smoke ring. A looped vortex passing through a field of photons would not be deflected by an electromagnetic field because a smoke ring vortex spins in opposing directions and can't climb the field like a spinning wheel moving forward. All this takes energy and the fuel driving the spin comes from the cancellation of (+ and -) ccwp's that randomly form in space.

Polarized light, on the other hand, is like a cross section of a donut—there will be two circling fields that make sign waves, one rotating opposite to the other as it moves through space like two wheels riding on one another through space. This line up of spins will make it possible for the polarized light to travel though a slotted opening like a key in a lock—molecules in a polarized lens do this. A polished mirrored surface of a magnet will cause polarized light, also a piece of paper will also due this as its surface has an electric field. Of course the light must reflect from the surface at the proper angle.

Electrostatics

Two like charges attract because (- and +) ccwp field charges cancel between them, while the outside [ambient] ccwp fields are packed together twice as much as between the particles. Impacts between like charges drive the particles apart as twice as many impacts occur between the two particles as is opposing impacts from outside. Energy comes from space as random movement of space chains vibrate by chance together to form a wave. This feeds the cancellations between unlike particles and also gives the particle core it's cube shape.

Fig 13

Quarks & Lepton Cores

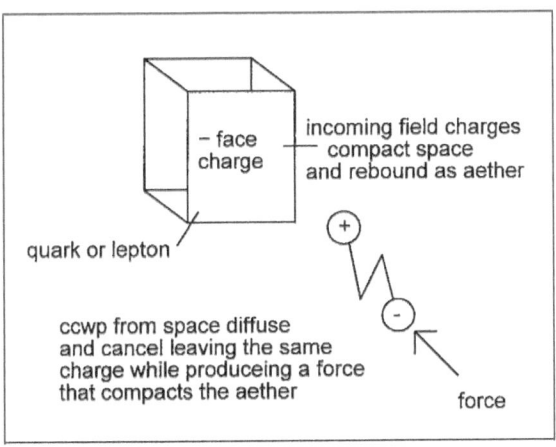

Space is filled with + and - charges randomly created from compaction and rarefaction of space . . . the charge at the face of the particle will cancel the opposite charge of + and - space charges and leave the original charge at the face of the core. The + and - ccwp's from space will diffuse into the space surrounding the face of the core. Canceled space charges form incoming compacted space particles that compact space or rarefied into the core and rebounds back into space, forming a field. For every vacuum particle wave created there will be a pressure particle created at the same time and this means all space will be 50% each particle. Note that the charge between the positive and negative field charge make a wave length during cancelation to the face charge in fig 13. This means the faces of quarks and leptons pulsates.

Like particles cancel each other therefore each face of the cube can have only one charge (+ or -). If all 6 faces of the cube is negative the electron will have a (-1) charge. This means that each face must have either a (+ or -) charge. Below is a list of quarks and leptons having these charges.

Chart

Showing Construction of Quarks and Leptons

face charges +	of face charges -	resulting charge	particle name
6/6	-0/6	= +1	positron
5/6	-1/6	= +2/3	up quark
4/6	- 2/6	= +1/3	down anti-quark
3/6	- 3/6	= 0	neutrino
2/6	-2/3	= -1/3	down quark
1/6	-5/6	= -2/3	up anti-quark
0/6	-6/6	= -1	electron

The cube is orientated to co-ordinates of aether space with face charges of (+ or -) fixed to a each side of the cube of matter particles.

Pauli's Exclusion Principle

Because ccwp probability waves impact the core are at right angles to the face of the core of a quark or lepton, the fields at the face of the core do not interfere with one another.

Relative energy of Quarks and Leptons

When particles bond, energy is releases as particles—the tighter the bond the more energy is released. Bonding between particles can also be induced by squeezing particles together. Mass is directly related to force and mass by the formula f = m@. If all faces of quarks and leptons have 6 face charges in which crossed chain wave particles + and - move into the charge at the face and become random ambient space there wouldn't be any reason to believe that all quarks and leptons would have the same mass. If this is so what causes mass?

Anti-particles have the same mass as particles do, charges have nothing to do with mass. There has to be particles bonding together to increase mass or in-line identity due to space compaction or rarefication. Another source of mass would be in the field of the particle. This is true for neutrino's which have a very low mass. The mass of a particle is due to compaction or rarefaction of space at the faces of the core. Mass can also be generated by capturing force particles such as light—a hotter object has more mass that a cold one.

Fig 14

Quark and Lepton Families

muon

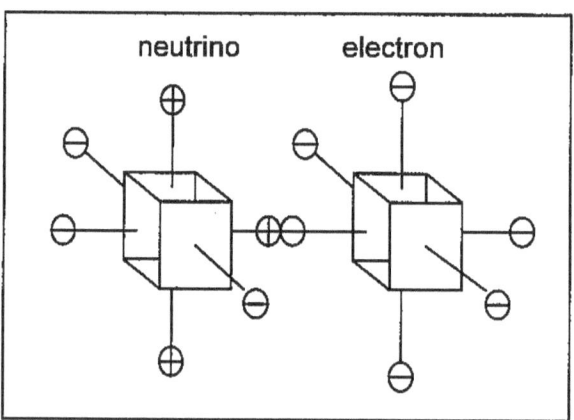

The three known families of quarks and leptons can be made by adding a neutrino to a side or sides of the quark or lepton core. In this case there would be 6 families of quarks and leptons or more.

Klien's paradox, field swapping

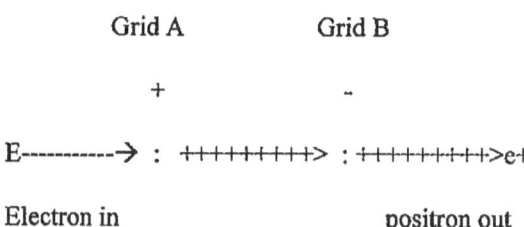

The drawing above is an experimental apparatus demonstrating Klien's paradox. If electrons pass through grid A at sufficient speed through a strong enough positive grid at A and then passes through a strong enough negative charge at grid A it will be come a positron.

Assuming that the electron has a compressed aether core with a field of bombarding ccwp, then when traveling through a positive field of more than double strength all the negative field will cancel leaving a positron having a positive field. The negative grid is weaker and draws the positron through grid B. It seems possible to change the charge of individual sides of the core to make other particles.

Using crossed chain wave theory it is easy to understand the conversion of fermions into other particles. The drawing doesn't show energy distribution or mass differences. The neutrino has very little mass while the electron has much more. The purpose of the drawing is only to show how charges of fermions can change when combining the face charges of particles.

Fig 15

Proton and Neutron Charge Configuration

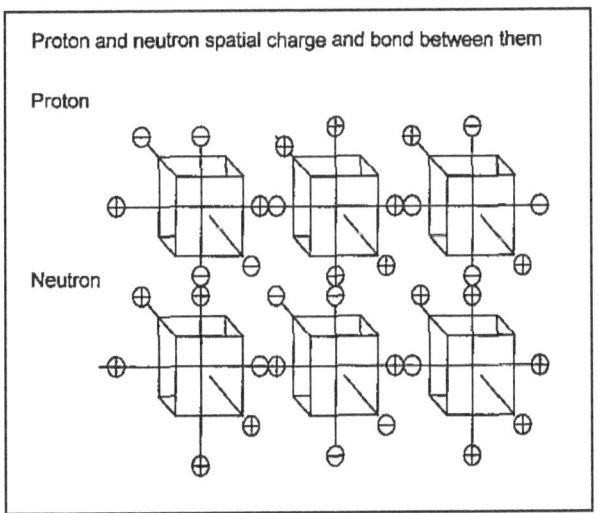

Protons and neutrons are fermions and are composed of U, P and S quarks. S stands for either and up or down quark in the structure. The upper drawing is an up-quark bonded to two down-quarks with a final charge of 0 and is a neutron. The bottom drawing is a down quark bonded to two up quarks and is a proton having a charge of +1. The charge bonding between quarks and leptons is the electro-w.

Fig 16

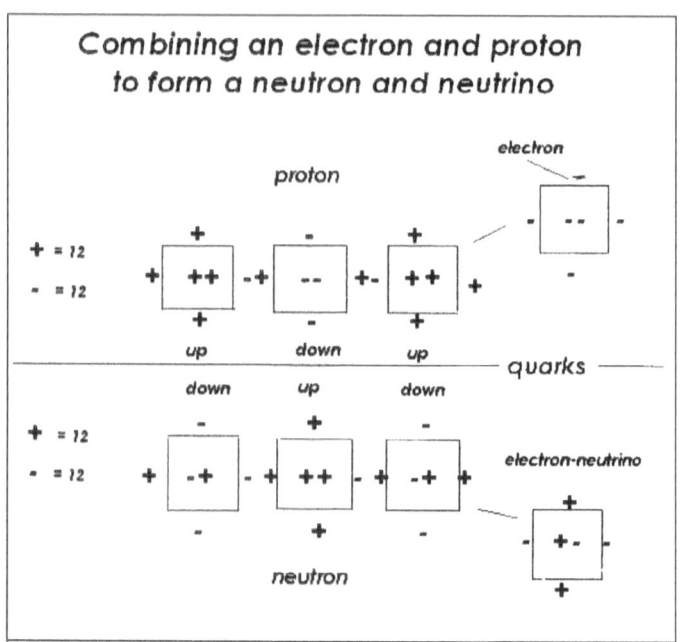

An electron forced into the nucleus of an atom would form a neutron and a neutrino as in a neutron star. The face charges of an up-quark bonded to a down-quark bonded to an up quark constitute a proton—bonding an electron to one of the two up-quarks causes the up quark to break its bond and exchange charges resulting in a down-quark that reattaches to the opposite side bonding with the now central up-quark making a neutron and a neutrino. The neutrino radiates out into space.

The reverse can also happen, a neutrino bonded at high energy into a neutron could result in an electron and neutron. Face charges can be changed as shown by experiment—an electron with enough momentum will penetrate a very strong positive field and become a positron upon emerging from the field.

Pauli's exclusion principle states that the three energy states cannot make a bond between two or three combinations. The spin, atomic number, and something else , one of the three must be different—the question is why.

The three quarks could be bonded at right angles instead of a straight line, however, the Pauli exclusion principle must be in the proper design. The strong bond is due to the close proximity in which cancellations between negative and positive charges cause vacuum between the quarks. A vortex spin is generated at each of the 8 apexes around the core of the quarks forming the fermion. The spins force ccwp fields outward weakening the strong nuclear force and allowing the quarks in the nucleus to flex or stretch.

At the core of the quarks there is a salient x-y-z spin as two opposing faces of the core move together then away. As one pair move together the other faces are squeezed outward in a churning sequence. In this way one quark in a fermion can bump another driving them momentarily apart. This is the cause of the weak force—the strong force is the very close electrical charges. In atoms, orbiting electrons will follow the salient two-sided x-y-z 3-step oscillations that generate a ½ spin. Plus and minus charges in ambient space are drawn by the + or - face charges of the core.

The loss of energy due to spin is energized by vacuum and pressure ccwp's moving into space and canceling vector movement into random movement of space. Random fields of ccwp's plus charge and negative charge reduce to random moving space chains leaving the ccwp's moving in the opposite direction to carry the wave forward. This action also compacts the core of quarks and leptons. I wonder how much time the neutral space regains its balance of + and - ccwp's after a wave travels through the space.

Types of Bonds

Electrical bonds are weak due to their distance from the nucleus of an atom. At the nuclear level positive charges of protons are close together and strongly repel each other, however, the negative and positive charges that occur in up or down quarks bond the nucleus in a stabilizing way. A quark will create an antiquark when a strong enough force and energy is applied to pull it away from a partner. The force holding the nucleus together is an impacting field photon force carrier in the dense aether around the nucleus. Stretched aether chains will align in one of the three directions of space orientation in a rarefied aether increasing in-line identity transfers causing mass—compacted aether space will also cause mass by increasing in-line identity transfers. In both cases movement is retarded. Positive charges retard movement while negative charges relax compaction of space increasing movement, but after canceling out the positive charge and moving into concentrated negative charge movement will slow down due to stretched ccwp chains and therefore mass will be created. In other words a negative field canceling a positive field will allow light to have a maximum speed through the aether.

Notice the two + charges at the left-hand side of the proton and neutron and the - and + charge at the right hand end. The nucleus must be constructed by these bonding charges and also the magnetic force of a vortex.

A Chart Showing Possible + and - Bonds Between Up and Down Quarks

Fig 17

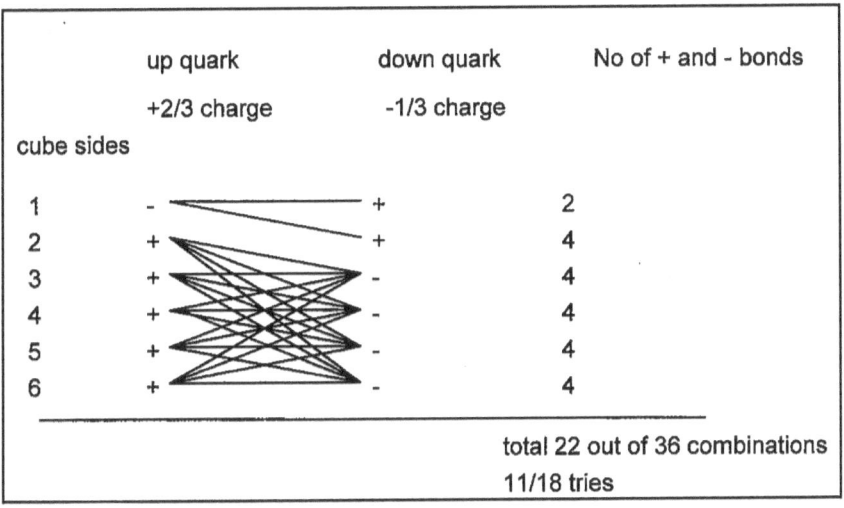

cube sides	up quark +2/3 charge	down quark -1/3 charge	No of + and - bonds
1	-	+	2
2	+	+	4
3	+	-	4
4	+	-	4
5	+	-	4
6	+	-	4

total 22 out of 36 combinations
11/18 tries

Color chromo dynamics uses colors red, blue and yellow to express the bonding of up and down quarks. One of the colors can bond with the other two charges, but like colors or charges cannot bond. This means the chance for a bond is 2/3. Ccwp theory, as shown above, is 11/18, however color bonding is 12/18. Only 1/18 difference. The problem is how accurate is the 2/3 color figure? If I'm right and color is wrong we don't need color to explain the bonds, they are just close (+ and -) 1/6 field charges fixed around a core.

Confinement between the quarks is thought to be a change in color chromo dynamic charge between the three quarks in protons and neutrons and mesons. Because of these color changes screening and camouflage around the particles the weak force is confined to a short distance.

In ccwp theory camouflage and screening are caused by field charges around a cube core of up or down quarks. The core is permanently orientated to aether space co-ordinates. Each core face charge is fixed around the core and can't be changed and the random pattern of face charges of quarks cancels the weak force in a short distance. Each face charge is 1/6 of a full particle charge.

Double slit experiment

When light passes through two properly spaced small slits in a piece of paper, a pattern of parallel lines can be seen on a screen. The reason given is that interference occurs between photon waves that cancels or reinforce. This experiment seemed to suggest that instantaneous-action-at a distance was occurring.

Ccwp theory explains the interference differently. Photons are surrounded by a field of space ccwp (+ or -) charge that pilot the photons by cancellation or bringing together like charges. When a photon passes through one hole the material between the slits absorbs some of the field causing a field collapse that drives the photon off its path causing an oscillation between the photon and its field. A single photon passing through the slit will cause a striped pattern on a screen. Ccwp theory suggests that in space some field ccwp's are traveling faster than the speed of light and pilot the photon.

Heisenberg's Uncertainty Principle

Time-leaps are part of "Heisenberg's Uncertainty Principle" . . . stating that if you know the position of particles you will not know the momentum. In between time-leaps there is a fixed rate time-continuum progressing until a quantum leap to occur in an electro-magnetic wave. The formula is delta T = Planks constant h divided by 2π dirac's constant. A pie distance means a 100-degree arc (sine wave) with a plus charge is completed, then a negative 100-degree arc circle to make a full circle. This could be explained with a vortex and could be the reason for ½ spin. A time-leap at the basic level of aether theory is not the same as Heisenberg's Uncertainty Principle.

The length of quantum-leap is equal to the number of rows in a crossed-chain-wave-particle that travels at the speed of light. 2π, in the equation, means that a distance of one turn of a wheel traveling on a surface is the distance of a quantum-leap in time.

Einstein's EPR experiments

Bell's inequities gave mathematical proof that Einstein's EPR experiments must be correct, however, experiments showed that Einstein was wrong. So it was assumed that between local events there is an instantaneous-action-at-a-distance to explain the facts.

The experiment goes like this; a polarized correlated light source emits horizontal H polarized light in one direction and vertical polarized light in the opposite direction. Analyzer H is positioned to sense H polarized photons and analyzer V is positioned to pick up V polarized photons.

Einstein's thought experiment assumed that when the V and H analyzers were 45-degrees angled from one another the correlations between the emitted V and H should have a value of $\frac{1}{2}$ polarization comparing the readings between both analyzers showing the number of photons having the same polarization between the two analyzers. The problem is that the experiment didn't give $\frac{1}{2}$ correlation between the two photons, but was $\frac{1}{2}^{\frac{1}{2}}$. It was then thought that each correlated photon knew what the other photon was doing and altered its condition.

The true measurement of correlation is the length of the chains in the component cross chain wave particle. There is as many time-leaps in each component, as in the photon which is made of many crossed chain wave particles. As crossed chains become closer to the length of ccwp composing light the greater the correlation limiting the options in the component. When the x-component length equals that of light there will be 100% correlation. In fact why call it correlation at all?

Scientists are moving analyzers farther and farther apart and find that they still get the same value—it persists at any distance. There isn't anything supernatural about this, its just in the correlated leap-time/row in a component chain . . . light has a given length of time-leaps/row and its component time-leaps represent the time-leaps used from light.

There will be as many time leaps moving in a photon as through the components. This means that there will be more smaller ccwp in a component than in the ccwp making light. At 45-degrees between analyzer A and analyzer B the length of a chain is $\frac{1}{2}^{\frac{1}{2}}$ and this is the correlation factor needed to correlate between analyzer A and analyzer B and a cosine function not a cosine square function is correct. Bell's inequities are wrong due to a faults premise. Even if both analyzers aren't aligned to space coordinates, the function is still a cosine function.

The paradox of star speeds at the outer ends of a galaxy

Stars at the outer reaches of a galaxy are traveling to fast to be in the orbit they are in. This can be explained by crossed chain wave theory. Crossed chain wave theory states that matter is compacted space and this means that further away from matter the less dense will be the chain aether. Less compaction of the space chain aether the less chance for in-line identity transfers and the faster light will travel. The space ship Voyager was further out in space than was anticipated because of the thinning of the space chain aether.

Fig 18

Correlation and crossed chain wave theory and hidden variables

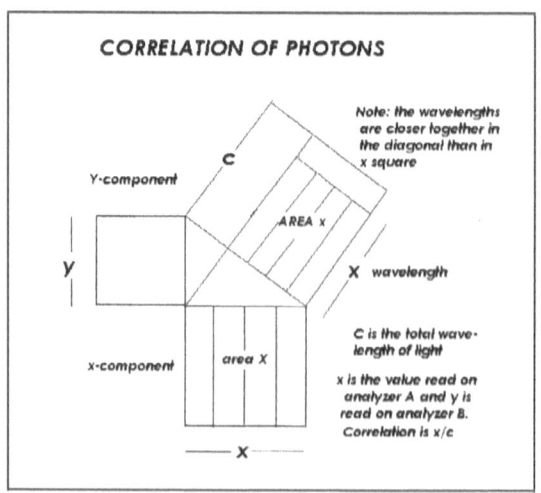

When a meson disintegrates, two photons are emitted, one moving in the opposite direction of the other—one photon is perpendicularly polarized as to the other. In ccwp theory correlation is related to ccwp wavelengths between analyzer A and analyzer B. Analyzer A will observe horizontal aligned particles while analyzer B will observe vertical aligned particles from correlated particles as correlated photons are perpendicularly aligned (correlated particles are 90-degrees apart in alignment so both analyzers are reading the same thing). The numbers of particles emitted will be registered and compared for correlation of vertical to horizontal readings. For instance at 45-degrees a center point for correlation is *1/2* of 45-degrees which is 22.5-degrees. The *1/2@* between analyzers gives the condition for equal wave lengths.

Note that the triangle (component) changes size from zero to the full wavelength of light that travels along an x, y or z space orientation—there isn't any true diagonals only time-leaps. The wave projections from the x-side of the x^2 rectangle to the projected side of the rectangle on c^2 are contracted while the true areas of the square and rectangle are the same, meaning that there are as many time-leaps in both areas. The wavelengths of the square are projected to the triangle, but do not fill the area on the rectangle. The projection from the x-square to the diagonal of the rectangle is equal to the wave length of C related to the wavelength of light.

Both analyzer A and analyzer B observe an equal sized wavelength of the square on the x-axis—one analyzer sees a horizontal wave pattern and the other a vertical wave pattern due to the opposite polarity of the correlated photons separations from each other. These chain segments, of a ccwp, form the square shown above. At 45-degrees there will be as many ccwp's for analyzer B as for A—however there will be $\frac{1}{2}^{\frac{1}{2}}$ that correlate There will be more smaller ccwp's in a Plank volume of light than would larger ccwp's (remember that the diagonal is a projection). Projections of unit measurements from real leaps are projected to line ab.

X lengths of sine waves from the x area fit into the x-area of the square of the hypotenuse that is a ratio x/c is the side of the larger square divided by its hypotenuse. To correlate with correlated particles between both analyzers must receive a photon that is part of the whole diagonal squared. At 45-degrees, correlation is $\frac{1}{2}^{\frac{1}{2}}$—this is the chance that a photon emitted will correspond with either analyzer A or analyzer B.

Space is filled with ccwp's and any Plank volume of space will have x-y-z components—only the x and y components for a photon are seen in the analyzers and these represent the x on the analyzers. This is because the analyzers are not aligned to space coordinate system like ccwps are.

A thing to remember, the emitted waves are correlated—this means that correlation will always exist if they align to the analyzers. All correlated particles are at right angles to one another. The larger the wavelength the greater the chance for correlation, because more of the wavelengths are in the larger particles.

A Plank volume picked has $1^2..2^2..3^2$ . . . .n^2—number of ccwp's in it. The larger the x component of light the greater the chance for correlation. Both analyzers are seeing the x-component and it will be 100% correlated the rest of the components are not. Each time different angles are presented from the emission source some will not be picked up by aether analyzer because it doesn't have the proper wavelength. Those that do are correlated as a cosine function and not a cosine square function.

Fig 19

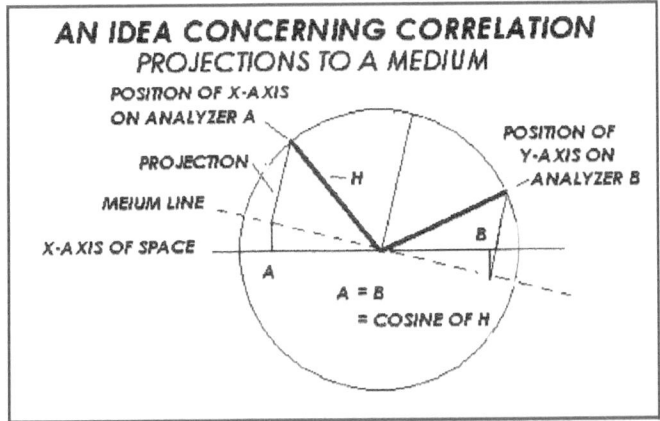

I came up with the idea of a medium line to represent a balance between analyzer A and analyzer B but was disappointed in the result because the medium line to the x-axis is a cosine function but from the medium line to analyzer A is another cosine function—the result is a cosine square function.

Fig 20

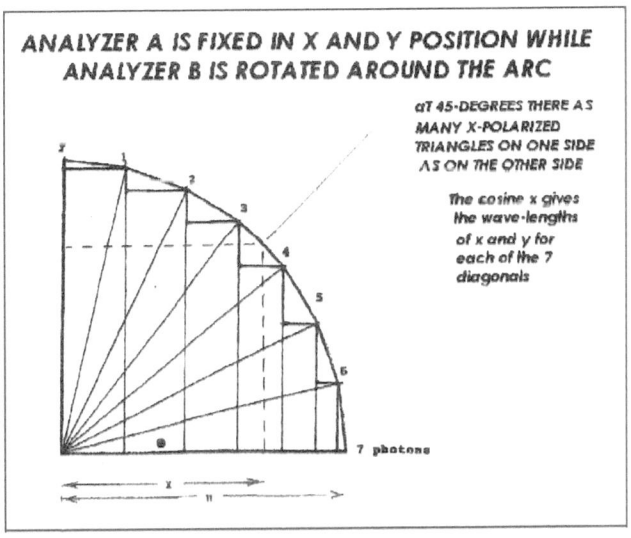

Figure 20 has 7 equally distributed ccwp's. X and y components are projected on a 90-degree arc. There are as many photons on one side as on the other at 45-degrees. The x-component of the ccwps that compose the photon show a projection of a cosine function. The x on an analyzer is not the x relative to the coordinate of space but will contain x and y components representing an x value.

Polarized light and crossed chain wave theory

Polarized light has transverse waves that have a fixed oscillation through space, while the angle of the transverse wave can be any angle relative to the space aether as it travels. The photon travels through space filled with (+ and -) field charges, meaning the photon must have a charge and would generate an electric field. However, most photons are neutral. I think a photon can come in (+ and -) pairs one following the other. Maybe, the oscillation is an action between the two.

Seven photons are placed on a ninety degree arc in the drawing. At 45-degrees there will be as many photons on one side as the other. The projection down the radial line to the x-axis at 45-degrees is 1/2, and this is the amount of light passing through the analyzer the other ½ is canceled out.

Twister theory

Penrose's "Twister theory" involves warping space by using strings to produce general relativity. Local patterns in space fit together to form integrating spaces similar to the principle use in semiconductors.

These migrating spaces are thought to be quarks and leptons. Using imaginary numbers as Einstein did in general relativity Roger Penrose showed that only 3 dimensions and time are needed with out other dimensions. Spin is relativistic due to harmonics of spin. At least this is what I understand. An interesting thing about twister theory is Roger Penrose's approach to homogeneity using exponents from formulas by James Clerk Maxwell. There should be a natural symmetry between the upper and lower exponents so that they balance, but the powers are off center as shown by the list below. I believe ccwp theory will correct this flaw.

Particle	Homogeneity
Graviton	-6
Photon +1	-4
Anti-neutrino +1/2	-3
0*	-2
Neutrino -1/2	0
Graviton -2	+2

Twister theory can explain how symmetry can be altered, but cross chain wave theory may also explain the asymmetry. In ccwp theory velocity C is the number of energy points/row in a basic ccwp. The energy in a ccwp that is the building block of a photon is C^2—the exponent added to the exponents given in the chart balances out.

Particle	Homogeneity
Graviton +2	-4
Photon H	-2
Anti-neutrino +1/2	-1
0	0
Neutrino -1/2	+1
Photon -1	+2
Graviton -2	+4

What was missing was the exponent 2 converting from a stepped system to a continuous system

Compaction of Space Before the Big Bang

Fig 21

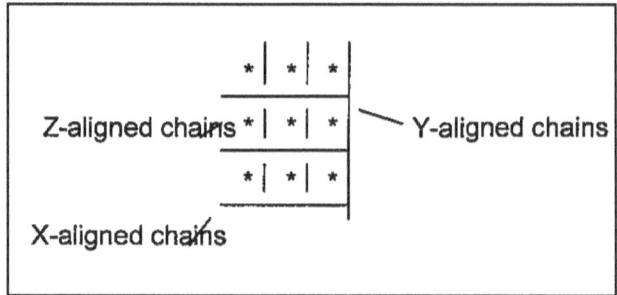

The drawing above shows a fully compacted space chain condition occurring before the big bang. All the X..Y..Z chains are completely compacted. The drawing is the XY view is of the face of aether chain compaction the * are Z directional taunt chains. The compaction of aether will only allow in-line identity transfers. This means time stops. Time begins as chains are released outward and other forms of identity transfers can occur. However, I don't think compaction of aether is this complete before the "big bang".

The opposite of the big bang is a chain loop square. Only in-line identity transfers can occur in this condition. Time stops but as the 4 sides of the square unravel time takes shape and particles can form, then life, and then the big bang. These cycles will repeat. I can't guess about the beginning or if it will end zillions of years from now.

Shiral Space

When the space aether is compacted, chains are drawn in the aligned as in the drawing above. The stacking of straight chains results in X-Y-Z alignments. Y-X-Z, Z-Y-X, and other arraignments are possible, but only one can exist at the same time. When moving in the Z-direction first x-aligned chains will occur, then y-aligned chains . . . this is repeated over and over, but when traveling in a diagonal to space you will encounter x-aligned chains, then y-aligned chains, then z-aligned chains forming a spiral through space.

Though not a true spiral it can stir crossed chain wave particles in space and produce a vortex that represents magnetism. The spiral path is reversed when moving in the opposite direction but is correct for the observer as he is facing that direction and sees the same rotation.

Fig 22

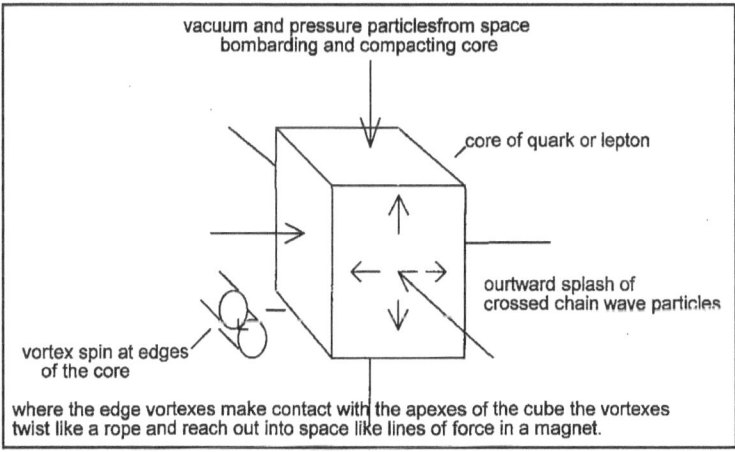

vacuum and pressure particlesfrom space
bombarding and compacting core

core of quark or lepton

ourtward splash of
crossed chain wave particles

vortex spin at edges
of the core

where the edge vortexes make contact with the apexes of the cube the vortexes
twist like a rope and reach out into space like lines of force in a magnet.

Fig 17 above shows how field (vacuum and pressure) ccwp from space impact space chains of a quark or lepton core resulting in an increase of inline time line steps slowing time which means an increase in mass.

The twin vortexes shown would occur when all faces have the same charge—when face charges are mixed then both faces would have the same spin and blend together as a single twist or spin. Sense the impacts at the core are quantum in force at the surface spin will have a quantum value for all cores of quarks or leptons.

Mass comes from compaction or stretching of space resulting in the same quantum mass for vacuum or pressure cores. It also should be noticed that the vacuum and pressure charges at the face of the core remain constant and equal numbers of ccwp's from space fall into the face of each core—positive charges will cancel linear movement of negative space ccwp's into random space movements (random movement of space chains propagate linear waves).

Each square face of the core will have a donut shaped vortex at its periphery and will have a vacuum at the center of the face if the face is negative and a pressure at the center of the face if it is positive—this also results in an opposite spin. This is the reason for charge at the face of the core. Monopoles do not exist in nature because they must draw energy to have any kind of life span.—the core of quarks or leptons get their energy from the chance of vibrating space chains that form crossed chain wave particles.

What is ½ spin?

I can't find a good reason for spin but here is my take on the subject. A normal torus will have a vacuum side and a pressure side and a complete circle around the torus will be at the start of the turn—this is a full spin. A half spin can be demonstrated by taking two pennies and fixing one coin in place while rotating the other coin around it. The rotating coin will take a full turn to position at the starting point and another full turn to end at the starting point atop of the fixed coin. The question is are we really dealing with true spin?

Suppose, instead of coins, we have a 4 magnets arranged around a central point having two north poles aligned oppositely from the center point and two south poles aligned oppositely across a central point. Take two of these magnetic arrangements so that, like the two coins, one is rotated around the other with a north pole face of one square to align with the south pole of the other square. Since the faces of the core of quarks or leptons are fixed in space and every other pole is opposite in polarity relocating the one square from top to the right hand side will again form a bond—this occurs in a half turn instead of a full turn. To an observer it would seem that, like the coins, one has rotated around the other a ½ turn—in reality there isn't any turn of the fixed position of the square faces of the core. It will seem that two turns as in the coins must occur to end at the starting point.

It seems that particle cores can be stacked into larger cores to give 1/2 spin—other arrangements could account for multiple spins. I don't feel satisfied with this concept though so I will look at other concepts.

The electric sign wave is perpendicular to the magnetic sign wave. They both start from zero to maximum value and back to zero in 1/2 the circumference then go negative in another half turn. Maybe this is the reason for 1/2 spin.

Fig 23

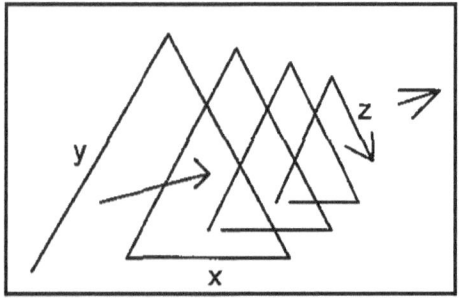

To move an object in a X,Y,Z diagonal of space a step must be taken first in the x-direction then in the y-direction and then in the Z direction. Of course a counter order can be made—however it would seem that there would be more stability if the spin is always in a given x,y,z order. A negative particle would have an opposite spin to a positive particle as negative ccwp's would have chains falling into the void as the void moves forward and the void is the negative particle.

Fig 24

Probability waves and atoms

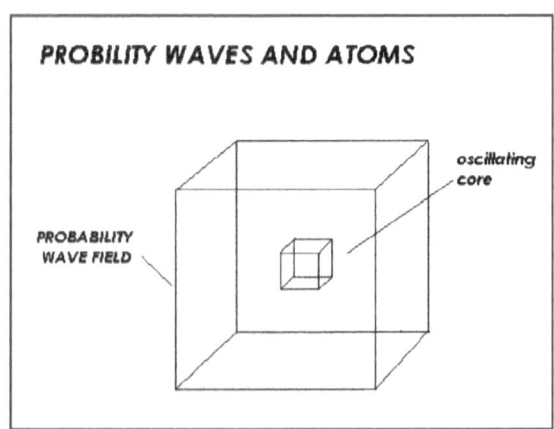

Probability waves would form rhythmical sequence of x,y,z planes allowing two opposite planes to move inward while the other opposite sides of the cube will move outward to produce a x,y,z spin pattern—this will allow chains room to splice into one another.

The two faces that move together will propagate by impacting crossed chains—chains that lie perpendicular to the planes are spaced between the crossed chains and don't intersect with the movement of the two clashing planes.

The core expands to the probability wave area shown and contracts back into a compacted cube of space chains. The next drawing shows extensions of the cubic probability wave intersecting each other and forming a cubic probability wave in space.

Because of interference between chains a sequence of x & y crossed chains must propagate in the z direction and then in the z & x crossed chains must propagate in the (+ or -) y-directions and then z & y crossed chains must propagate in the (+ and -) x-directions to make a vortex.

Electrons, moving in opposite direction will be drawn together by this spinning field while like charged particles are repelled—this is the opposite action to magnetic attraction and repulsion.

I particularly like this idea because the phase speed of orbits around these quarks or leptons would be constant and this means the speed of increasing orbits is the square of the distance—this would give the spins for each orbit.

Fig 25

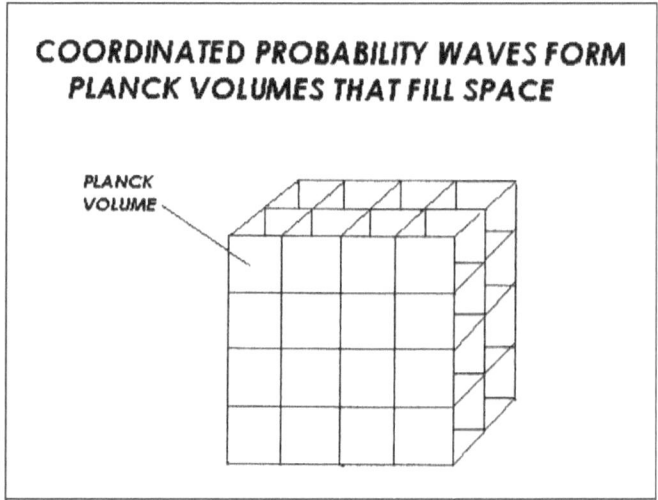

If the core is positive it will be surrounded by positively charged probability cells of diminishing charge. A proton will occupy a cell while an electron bonding with the proton will take up a cell position nearest to the protons cell.

(+ or -) charge fixed in a spatial order. Ccwp, being much smaller than quarks or leptons, form probability waves around these particles. Because ccwp can only travel in one of 6 directions, 8 boxes made of probability waves will form the first orbit which is a square around the nucleus of an atom.

There isn't any sensible energy in probability waves because a ccwp moving one way cancels the movement of another, like air molecules do. Ccwp are made of aether chains and these chains will flatten into walls that stretch out into space forming the space boxes. An electron can fit at the intercepts of these boxes as long as the nucleus can attract them and repulsion doesn't drive them out of the probability orbit. This means that the first orbit can have an electron in the center of each face of the larger cube orbit. The next orbit will have 4 cubes of 8 blocks this allows 4 electrons on one face and another 4 on the other 4 cubes (8 electrons in the second orbit. The third orbit will have 9 cubes of blocks and will have places for 9 electrons on one side of the orbital cube and 9 one the opposite side. This is true for all following orbits.

How the Core is Made

Here is a concept that might explain spin—first, two waves of crossed chain wave particles aligned to the x-plane collide then two waves of the y-aligned plain collide then two waves of z-aligned plains come collide in a repeating cycle. While one set of planes are moving out to the probability wave another set is moving to the core. This is shown by the formula for orbits $2n^2$, The 2 stands for 2 waves the n^2 stands for the number of crossed chain wave particles in the wave area if space. The x-y-z cycle generates a cube. Each probability wave will have a lower concentration of [+ or -] crossed chain wave particles in it.

Point locations between time-leaps occur continuously resulting in random movement of every segment of the point loop universe. Chance movements of the oscillating chains results in very small crossed chain linear wave propagations (random to linear). Probable forward movement of space chains results in a vacuum in which chains fall backward into the void while the void moves in the opposite direction. In other words when a pressure crossed chain wave particle is formed a vacuum (void) particle will also form and travel in the opposite direction.

A vacuum and a pressure particle will cancel each other back into ambient space and while they are short lived they are the building blocks of real and virtual particles. Cancellation of vacuum and pressure ccwp's is the cause of compaction of the core of quarks and leptons. The face charges of the core cancel opposite charges of plus and minus field particles leaving the original charge.

Fig 26

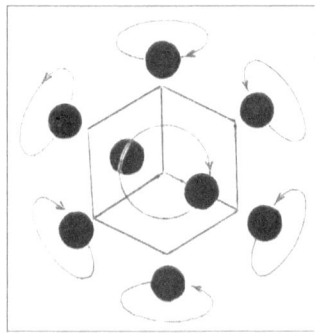

SPIN AND ELECTRIC FIELDS

The drawing above is about the cubic layout of a Planck volume of space representing a quark or lepton or any particle having a field charge. The balls that circle the space core, shown by the cube in the drawing, is Planck space filled with vacuum and pressure ccwp's of equal magnitude and therefore is neutral in charge. However, space resonance will cause the separation of charges into waves of ccwp's that impact the core—first a pressure wave will impact and rebound from the x-face and then a pressure wave will follow changing the face from Positive to negative. This action occurs at opposite sides of the Planck space core. Each of the three pairs of faces will pulse this way from the x-pair to the y-pair to the z-pair, The dragged space (balls of space) is drawn by the x, y, z pulsing face charge and forms a vortex at each of the eight apexes. The ccwp's are the alternating electric charge at the faces of the of the space core.

Each space cube has ccwp's that fill a Planck volume of space and travel in (+ and -) direction and form the cube cores of space.

A basic photon is a Planck volume of space and contains all the ccwp's N^2 components to give the speed of light in any direction to a photon. Sense opposite faces of the space cube move outward and clash with negative waves from the shock wave they cancel into random space while the components that move in the opposite direction continue the process through space.

Quarks and leptons have a cube core just like a Planck cubic volume of space but have only one charge (vacuum or pressure) at each core-field charges will also be dragged by the x,y,z spin of the apexes of the **core.**

Up-quarks and down-quarks make the nucleus and electrons will follow the induced spin of the space field. The orbits comply with electron repulsion and orbital induced spin.

The pulse rate of the phase spin will cause electrons to move faster in orbits that are further out and could account for the greater magnetic field or spin characteristics of the electrons.

Fig 27

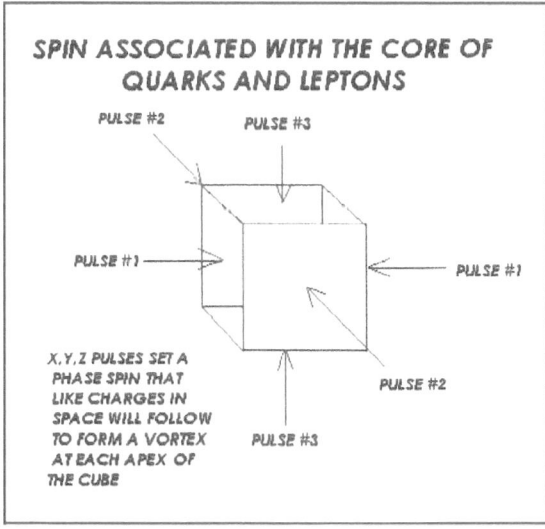

The x-y-z spin sequence stirs the field of photons into a half spin. Can this type of oscillation really exist?

Also this doesn't explain all the spin possibilities from zero to *1/2* to 1 to 3/2 to 2 and so on. The real question is what does spin mean? Does a half spin mean that two turns around a particle such as an electron will return to the starting polarity? Electrons do pair as in a covalent bond between atoms—the action is like two monopoles drawn together by fluid dynamics. The electrons would be fixed to coordinates of space so that moving around the electron would simulate a full circle around one electron and then the field is shifted so that another full circle around the other electron would bring it back to the starting point.

Atomic Orbits

The formula for electron orbits is approximately $2N^2$. The formula seems to suggest two planes coming together and rebounding to form probability waves while spin at the nucleus of an atom is another factor.

---First plane
................*.......*   Fourth Orbit--16 electron
.....*.........*.......*   Third Orbit----- 9 electrons
..........*.......*   Second Orbit -----4 electrons
............*   First Orbit -- I electron

The nucleus is located here.—Second plane below

............*   First Orbit -- I electron
..........*.......*   Second Orbit -----4 electrons
.....*.........*.......*   Third Orbit----- 9 electrons
................*.......*   Fourth Orbit--16 electrons

2 x-planes come together with the above configuration then retreat as 2 y-planes come together in the same configuration and retreat as 2 z-planes come together in this rotating sequence. Electrons follow the sequence forming an x-y-z spin. N^2 number of electrons is caused by probability waves boxes shown in fig 14. Electrons can be found at a corner between 8 boxes.

Other forces such as cancellations and diffusion flow of + and - ccwp's and aether flow cause a vortex that alters the basic concept pattern. Fluid dynamics theory is fundamental to a complete understanding.

Note that an electron will have to follow a salient x-y-z phase spin. An electron has a half spin but in an atom it has a spin of one. The spin of an electron is twice as much as it should be. An electron moving at the speed of light wouldn't have it's spin—it would be converted into forward movement. Lines of force from bombarding space charges that cancel into random space at the face of the quark or lepton would form a toris at each face of the core of quarks or leptons and a vortex reaching out into space at the 8 apexes.

The orbits above though roughly correct are not completely correct due to vortexes and other forces in an atom. The x, y, z phased oscillation is constant and therefore each orbit will double its speed and this contributes to the magnitude of the spin. The first orbit would have 1/2 spin, the second would be 3/2 the third orbit would have 5/2 spin—each spin is increased by its diameter plus 1/2. If the first orbit had a spin of 1 the second orbit would have twice the spin and the third orbit would have three times the spin as the first and so on. Assigning a half spin for the first orbit spin means that each orbit must have a half spin subtracted from the relative diameter of its orbits.

Fig 28

Space Shirality along diagonals

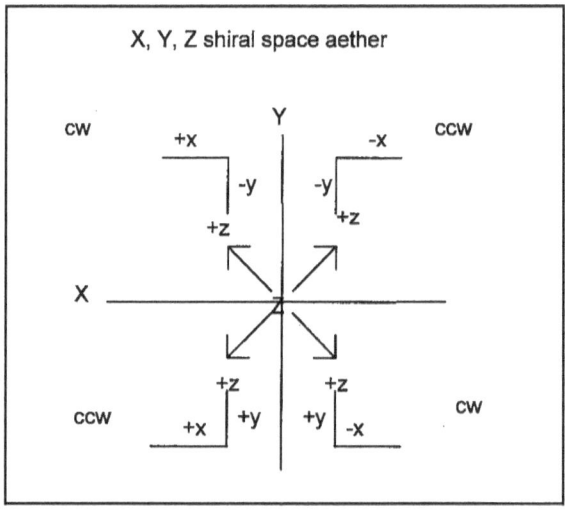

The drawing above shows the back side of a room or the backside of a core with spin in the observers direction. If space is laid out in x,y,z fashion then spin will be as shown in the diagram. An incoming wave will travel along the x-axis then along the y-axis and then along the z-axis.

Fig 29

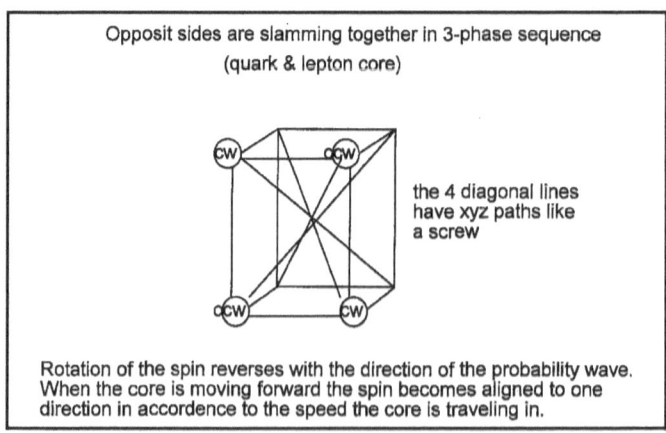

The drawing above shows the x-y-z spiral from the face shown to an opposite comer much like the thread of a screw . . . there will be 4 of these two will have a left hand spin while the other two will have a right hand spin. Spin is due to the x-y-z layout of space. Its not clear if this type of spin does happen.

Fig 30

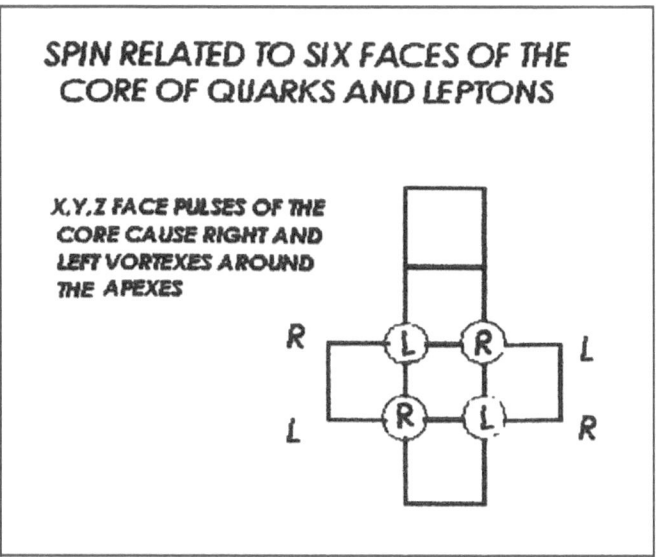

1) X,Y,Z spin in any coordinate system will be the same for all particles—therefore relatively no angular spin will happen between them—unless one particle bangs into another.

2) All particles must spin in the same direction or they would interfere with each other. In other words a x to y to z movement of all the particles in the mass follow a spin along the x,y,z diagonal of space.

The drawing above is a core of a quark or lepton. To move from start to the opposite side of the cube the field must step along the y-direction, then in the z-direction and then in the x - direction around the edges of the cube, then the return will have a -z, -y, -x return spin. From there follow the line x, z, y to the opposite edge and then return -z, -y, -x. It will take 2 turns to end at the same starting point.

Sense diagonal movement around the core of a quark or lepton is x,y,z steps there will be spin between the opposite apexes of the core 180-degree rotation with a positive charge and 180-degrees with a negative charge. I wonder if a half spin represent a 360-degree turn having two diameters of travel. Each ½ turn makes a full spin and a return makes a full spin just like moving a coin around a like coin that is fixed in position the one moving around will take 2 turns to rotate around the fixed coin. I see this happening with lines of force surrounding two particles one moving around a fixed particle. The line of force will experience a rotation like the coin since the core faces are fixed in space.

Actually this doesn't seem to be the case for ½ spin because it takes two turns around an electron to arrive at the starting point. I don't know what this means—does it mean that two poles are attracting only after two turns? I can visualize this happening with two gears—the fixed gear would be an electron while the rotating gear would be 1/2 the size of the fixed gear. After one trip around the fixed gear what is the condition that makes it necessary to take another turn to end at the starting location? Is there something special about the starting cog—for instance south to north pole. It is known that spin can be flipped but is there 6 fixed positions in space for spin as around the face of a cube? If spin is fixed in space it would be quantized as on the faces of a cube—if not it couldn't be quantized. OK, something is special about the fixed gear, but why? An observer moving a field around an electron twice must observe something that makes him think he has returned to the starting point.

Spin is also relativistic and a Plank volume. A photon is a Plank volume in which n^2 where n is I,2,3,4-to the maximum size a ccwp can be. The drawing below shows how spin is time relativistic when an electron circling an atom is driven forward at the speed of light. It will take 2 turns rather than I because rotation in time is diminished. We do know that time slows when an object picks up speed.

Probability waves around photons

In ccwp theory Photons are composed of ccwp that average the speed of light. This means that some of these ccwp will move faster than the speed of light and should cause probability waves relative to the photon. By averaging the others that move faster than the speed of light we could find the wavelength of the probability wave relative to the photon. The speed of light may propagate from zero to 2c averaging the speed of light.

The same number of the ccwp field particles move ahead and back. To achieve the forward speed a ccwp will be smaller but will be more of them while the return will be larger and slower with less of them. To achieve the speed of light C it must have p^2 number of points.

Ccwp theory states that in a Planck volume there will be ccwp with sizes $1^2..2^2..3^2 \ldots N^2$ particles. $N^2/2$ of the particles move faster than the speed of light the other half move slower than the speed of light. However, ccwp that move slower than the speed of light are larger than ccwp that move faster than the speed of light. Therefore, fewer particles move forward than in the reverse direction. The Planck volume is divided between particles that move faster than the speed of light and those that don't.

On the other hand, in a probability wave, as many points move in one direction as in the opposite direction. When moving forward, ccwp are of smaller size, and fewer larger ccwp when moving in the opposite direction. To have the same number of energy points more smaller ccwp will move in the forward direction and fewer larger moving in the opposite direction. This can be shown by the formula $C(xyz) = x(Cyz) + y(Cxz) + z(Cxy)$. This concept could work at sub atomic levels but light propagates through an aether as ccwp's of C^2 size.

Explanation for the Bulls eye pattern in the "double slot" experiment

A fluid aether can best explain the bull's eye pattern on the "double slot" experiment and at the same time include spin as the makeup of a photon. This includes a vortex forming one part of the photon and a counter rotation vortex as the other part of the vortex. The bond is strong but can be broken when passing through one slot . . . the other slot can include the counter rotating part of the photon. Interference comes when the two parts reform due to decreased pressure between counter rotational vortexes of the two parts of the photon.

Another explanation was given by de Broglie for the bull's eye pattern in the "double slot" experiment . . . pilot waves that direct the path of the photons . . . ccwp theory gives an explanation for the pilot waves. Pilot waves form probability waves that in turn intersect and form the bulls eye pattern.

For a photon to travel through both slots, the photon must either have two parts, one negative and the other positive, that separate enough so that one part of the photon can move through one slot while its partner moves through the other slot, or its pilot wave moves through both slots.

Instantaneous action-at-a-distance for gravity between gravitational systems

Gravity between distant objects is instantaneous, it is not a reaction occurring at the speed of light and back., experiments have proved this to be so. Ccwp has a built in explanation for this effect. Each cycle of the chain aether path balances all points in the system. This means that the gravitational field moves with the object having the field. I don't believe a gravitational signal travels from an object back and forth between objects at the speed of light. It's all in the compaction or rarefaction of space chains of the aether. Gravitational action is local.

Aether Density and Size and Speed of Particles

The speed of a particle such as a photon would be higher in a thinner aether chain media because aether chains would have less resistant to motion with less in-line identity transfers. I believe the aether is physical and carries all the energy of propagation as air molecules do. All movement must start in a time-line, one step following another around a universal looped path in x-y-z steps. In this way every movement is in step with the rest of the universe—if it weren't, time would change randomly through out space.

In the step movement condition, waves could form a universal balled path. Random movements would, by chance, form linear waves that could form photons, that could form virtual fermions, that could form vortexes like water molecules can. The vortexes could form all the particles and forces in nature.

If waves can form randomly in space, they can also dissipate into random movement of the stepped aether loop. This leads to conditions where lost energy, in photon travel, could retain its energy. For instance, if a unit of space (Planck volume), by chance, has the proper energy to continue propagation of a photon, it will do so one Planck volume at a time. If I am right about this it would take time for space to return random movement of space chains into vector waves of + and - ccwp's.

Kaluza and Klein's 5th dimension

Kaluza & Klein showed that another dimension is added to the four dimensions of relativity a relationship to gravity becomes apparent—if one of the dimensions is curled up electro-magnetism resulted. Electromagnetism occurs when an object is moved at the speed of light and becomes a wave in one direction—of course all lateral wave movement isn't stopped, light has lateral movement or polarized light couldn't happen.

In ccwp theory there are two types of ccwp's, one is reverse bias to the other. A positive ccwp has chains that move forward while negative ccwp's have chains falling into the void and compacting as the void travels forward.

An electron is a particle made of negative ccwp's and as it propagates through space, particles that are made of pressure particles and vacuum particles that spin around each other make neutral fermions that fill space and rebound off one another. The path of the electron, as it travels, cancels opposite particles into random space that the electron propagates through leaving its original charge. This causes space charges to rain down on the path generating a force at right angles to the travel path. Diffusion of the particles raining down is the force generating magnetism. The canceled charge becomes the aether in which diffused particles can propagate through.

In ccwp theory there are only 3-dimensions or steps in x-y-z directions. Inline identity transfers result in gravity—or stopping movement of the path. If all the $x^2 + y^2 + z^2$ components for lights are vectored in a given direction to equal c^2 local time stops as in relativity. To accelerate an object to the speed of light an infinite amount of matter is needed to do so—this increases gravity. Increasing mass also increases inline identity transfers and thus increases gravity.

Kaluze Klien's 5th dimension is gravity and subtracting a dimension results in electromagnetism. What dimension is removed? Suppose the 5th dimension is inline identity transfers in relative gravity to curl up relative mass or space denseness electro-magnetism. Electromagnetism is a movement relative to an observer. Which one of the dimensions is curled up? Mass is due to crossed chain identity transfers and inline identity transfers both are dimensions. Forward movement slows time due to vector change and an increase in mass. For instance if all the time-leaps are in one direction there won't be any leaps in other directions, this leads to a time curl. What about x-y-z spin? To an observer opposite spins cancel as in a stationary electron but backward spins cancel leaving forward spin. Does this account for ½ spin? In an atom electron spin is twice as high. This Is due to aether + and - ccwp's space following the pulses of the nucleus.

Dark energy and dark matter

Dark matter and energy are the same thing, a compaction or rarefaction stretching of space chains causing in-line identity transfers is the cause of gravity. The ratio 3 to 1 of energy to matter in our universe has been explained by Dr. Hong Sheng Zhao of the University of St. Andrews of the school of physics and astronomy as being caused by the expansion of space. This makes both matter and energy the same thing.

If ambient space has an area of lower density, light will travel faster than in the ambient space—if an area of space is compacted light will travel slower though it. This means there will be two types of dark energy and matter. If the lower density dark matter and higher density are brought together a large black hole should follow as space chains rush in to fill the void. It doesn't seem that energy would be lost here due to cancellations. The uneven expansion of space would cause equal amounts of the pressure and vacuum of dark energy and matter.

How to curl dimensions

1) Inline identities generated in the uneven packing after the big bang leads to vacuum and pressure spots in the aether. This leads to a time delay in the compacted space area and a time increase in the vacuum spots of space—curled space.

2) Ccwp's slow time as time-leaps across chains of a ccwp are enlarged—another curl in space.

3) Relativity slows time when an object travels through the aether as the speed of waves is relative to the aether or fixed space.

4) Expanding space will cause the speed of light to travel several times faster than the speed of light relative to a fixed point in space. The time-line aether moves much faster than the speed of light and carries the propagation of light so that it travels much faster than the speed of light.

Then time is slowed it is curled as though it were in another dimension. Fixed space cannot be compacted and is actual distance while aether in space can be contracted in distance and speed of movement are part of curling in 3-dimensions.

Conclusion

So far the theory seems to explain the three dimensions, time and everything from a simple beginning to forces and charges to gravity and the big bang and the field charges of quarks and leptons. Also, it explains instantaneous—action-at-a distance, how energy is extracted or returned, by chance from the fluid ether, how the double-slit experiment has a cosine function, the reason for pilot waves, the relation of a Plank volume of space to a Plank particle, and what a photon or light wave is. A lot of my concepts could be wrong but my hope is that it could inspire thinking that leads to a real answer to a theory of everything.

The thought process involved in this theory is the same as evolution theory, an observation and conclusion of facts. Everything in the theory fits together by an evolutionary process that becomes more complicated in time. Like Darwin's evolution theory, scrutiny will have to work on the theory for years and years with new proofs. A theory must be able to predict events and so I will predict that, in time, ccwp will be found and they will have a plus or minus charge and travel in 6 perpendicular directions. Of course this could be fiction not science.

Gerald W. Siegle

www.ingramcontent.com/pod-product-compliance
Lightning Source LLC
Chambersburg PA
CBHW022115170526
45157CB00004B/1658